Starting with Bees

Peter Gordon

Broad Leys Publishing Ltd

Starting with Bees

Published by Broad Leys Publishing Ltd: 2004, 2007

Editor: Katie Thear

A catalogue record for this book is available from the British Library.

ISBN: 978 0 9061 3735 2

Outside front cover photograph: Anna Chambers

Outside back cover photographs: Katie Thear

The following picture sources within the book are gratefully acknowledged:

Anna Chambers, Katie Thear, E H Thorne (Beehives) Ltd,
ClipArt Collection, and DEFRA.

Thanks are also due to the Saffron Walden branch of the Essex Beekeepers'
Association for allowing themselves to be photographed at a local apiary.

For details of other publications please contact the publishers:

Broad Leys Publishing Ltd
1 Tenterfields,
Newport, Saffron Walden,
Essex CB11 3UW, UK.
Tel/Fax: 01799 541065
E-mail: kdthear@btinternet.com
Website: www.blpbooks.co.uk

Contents

Preface   4

Introduction   5

About the Honey Bee   9

The Hive   25

Getting Started   41

Handling and Checking   46

Feeding   52

Swarming   55

Re-Queening   61

Moving Hives   65

Pests and Diseases   67

Honey and Other Products   73

Bees at the Show   81

The Bee Garden   83

The Beekeeper's Year   86

Appendix I: Bee stings   90

Appendix II: Legislation   91

Glossary   92

Reference section   94

Index   95

Preface

*For he on honeydew hath fed
And drunk the milk of Paradise.*

(Kubla Khan. Coleridge.)

I can't remember a time when bees did not figure in my childhood. There were hives in the orchard, pots of honey on the table, and the smell of bee flowers in the garden. Despite the occasional sting, I determined that I, too, would keep bees one day. There was a career to be followed, of course, but always at the back of my mind was that image of a garden full of flowers, an orchard and a 'bee-loud glade'. It was only after I married that I was finally able to have a couple of hives, initially in a suburban garden. It was quite a while before we were able to afford a place with a bigger garden and an orchard, although the latter was a tiny one. I have been a beekeeper now for many years, but have never lost that sense of wonder at the complex and interesting world of bees.

I am grateful to Katie Thear who has been a source of inspiration for this book. She has done an excellent job of commissioning the work, editing the manuscript, making suggestions for its improvement and presentation, as well as taking photographs to add to my collection.

Anna Chambers, too, has provided some excellent photographs which open windows into the beekeeping world.

Finally, my thanks are due to John Rayner, Membership Secretary of the Cambridgeshire Beekeepers' Association, for reading the manuscript and pointing out where improvements and amendments were needed in the text. An experienced beekeeper, he has been involved with the practical side of teaching beginners at the association's training apiary, as well as acting as Honey Show Secretary.

This book is written for those who are quite new to beekeeping. I'm aware that there are already many first-rate books on the subject, but many assume a certain level of knowledge and experience on the part of the reader. Here, no assumptions are made, and the reader is introduced to each topic on a step-by-step basis. However, I must emphasise that a small book can only provide an overview of the tasks involved. Only practical tuition through a local organisation can provide the wealth of detail necessary for successful beekeeping.

I hope that in a small way, this book will prove to be useful to those who are thinking of keeping bees, and that they learn to love the world of bees, as I do.

(Peter Gordon)

Introduction

Bees are not solitary creatures like eagles, but gregarious as are men.

(Varro. Rerum Rusticarum. 36BC)

Why keep bees? This is a question that I have often been asked over the years, and it seems to me that it is a good starting point for anyone who is thinking of beekeeping.

I like bees, and that in itself, is a sound reason for keeping them, but if there is anyone in the family who is allergic to bee stings, then it may not be appropriate to keep them.

They are full of interest, and have an amazingly complex hierarchy. Although I have kept bees for over thirty years, I am still learning about them, for the simple reason that there is always something new to find out about the way they function.

Relative to their size, the harvest that bees produce is greater than that of any domestic livestock. We all benefit from the availability of honey and beeswax, and the many products that are produced from them, or which include them as important ingredients. The list is long and impressive: honeycomb, potted honey, royal jelly, confectionery, food supplements, medicines, mead, candles, furniture polish, soap and skin care products.

Then, there is their work in pollination. Where would our orchards and kitchen gardens be without the bees that carry pollen from plant to plant, enabling pollination and subsequent fruit production to take place? Yes, there are other pollinating insects, of course, but they would be hard put to equal the amount of work that the bees do. Commercial fruit orchards rely on bee activity, and often have contracts with beekeepers to move their hives there, to maximise the level of pollination.

In ordinary gardens, too, long and lazy summer days would not be the same without the soft drone of bees working the blossom.

Are there any restrictions?
Anyone may keep bees, as long as they are not a 'public nuisance'. In other words, the hives should be placed in such a way that the flight path of the bees does not impinge on where members of the public may be walking. When bees emerge from the hive, they fly upwards at an angle. Then they fly off to their required destination, which in the case of worker bees, is to collect pollen, nectar, propolis and water.

Are Bees for You?

If you are interested in bees, but undecided about keeping them, the following may be useful in arriving at a decision:

For

Anyone may keep bees and there is no need to register.

They are full of interest.

Their pollinating activities ensure good fruit and other harvests.

They provide honey and wax which can be used in a wide range of products.

Most of the time, they require little management.

There is an excellent network of bee societies where help and information are readily available.

Second hand equipment is available (but it must be thoroughly cleaned to prevent disease transference).

Plans are available for making your own hives, if you are proficient at DIY.

Beekeeping is possible in urban areas where parks and gardens provide readily accessible flowers.

Compared with many animals and birds, bees have fewer disease problems.

There are many beekeeping courses available.

Beekeepers are said to live to a ripe old age!

Against

They must not constitute a 'public nuisance'.

They may leave marks on washing on the line.

A family member allergic to bee stings might not appreciate them in the garden.

At certain times of the year, they need extra management, eg, swarm prevention, feeding, harvesting the honey, etc.

If they swarm and leave your land, they can be taken by anyone.

You need at least two hives.

Setting up costs of hives and equipment are relatively high.

Beekeeping can be difficult in areas of intensive farming where there is a lack of forage.

Honey separation can be difficult in arable areas where oilseed rape is grown, because it sets quickly in the comb.

Disease prevention legislation must be followed.

Legislation controlling the sale of products must be adhered to, if sales take place.

There's a lot to learn about bees, although it's not as difficult as some make out.

It will be seen, therefore, that it is important to keep away from the front of the hive until the flight path is above the heads of people. Having a relatively high hedge or windbreak screen not only provides shelter for the bees, but also has the effect of making their flight path steeper.

It is interesting to note that one of the few pieces of legislation that applies to bees goes right back to Roman times. The principle of *ferae naturae* states that the beekeeper owns the bees while they are on his property, but if they swarm, they belong to whoever takes the swarm.

The only other areas where legislation applies are with disease control and the sale of produce. The former is covered in the section on *Pests and Diseases*, while the latter is to be found in *Appendix II*.

Where to start

Once the decision is taken to go ahead with bees, it's worth estimating how much time is involved, and where to get help and information.

Most of the time, the bees can be left to their own devices, with just the occasional check to make sure that all is well. For example, from April to August, during the beekeeping season, about 30 minutes per hive, a week, is sufficient. Extra time is required where activities such as making up and cleaning equipment, swarm prevention, feeding and honey separation are concerned. Holidays may need to be organised around these periods.

This book provides all the basic information about starting with bees, but it is also a good idea to gain some practical experience first. There are many excellent courses on beekeeping, some arranged by the *British Beekeepers' Association* (BBKA) and local beekeeping organisations. There are also courses held at agricultural colleges, as well as those organised by suppliers of beekeeping equipment. Before buying any equipment and setting up a hive, it is an excellent idea to attend a beginner's course and talk to those who already keep bees.

Join the local beekeeper's group and attend their meetings. They will usually have an interesting programme of talks by visiting speakers, and there will be members who are more than willing to help a newcomer. Help and advice from local beekeepers is invaluable. Not only are they experienced, but they are also familiar with local conditions. Many local associations also have equipment such as extractors that can be borrowed, to help keep the costs down in the first few years. Beekeeping clubs will also put you in touch with the local Bee Officer of DEFRA (*Department of the Environment, Food and Rural Affairs*). If necessary, he will carry out regular inspections of your hives, and advise on disease prevention, as well as what to do if there is a problem.

Beekeeping through the Ages

The earliest representations of bee colonies being raided for honey are in Paleolithic cave paintings in Spain.

(Larousse)

Honey was initially taken from wild bees wherever they happened to have made their colonies.

Aristaeus, the beemaster of ancient Greece. All the ancient civilisations were familiar with the principles of beekeeping.

In medieval times honey was still the only sweetening agent available, and bees were an essential part of the economy.
(British Museum)

Straw skeps were widely used as hives, but the bees had to be destroyed in order to get the honey.

The 'bee space' was one of the discoveries that revolutionised beekeeping because it meant that frames could be moved from the hive. It was no longer necessary to kill off the bees in order to get the honey.

The WBC hive became synonymous with the country garden, and is still in use today, although most beekeepers now use National hives.

About the Honey Bee

A land flowing with milk and honey.
(The Bible)

A brief history of beekeeping

We have no way of knowing how early man discovered bees and honey. Perhaps he saw bears helping themselves from the wild colonies in holes on trees, and thought that he, too, could benefit. The earliest representations of colonies being raided in this way are in Paleolithic cave paintings, so the relationship between man and bees goes back a long way.

All the ancient civilisations were familiar with bees, including the Sumerians, Greeks, Egyptians and Romans, and of course, there are also references in the old Testament. There is even an Aesop fable about bees! The Greek slave recounts how some worker bees had built a comb in the trunk of a hollow oak tree, only to have some drones claiming that it was theirs. The case was taken to the wasp for judgement. He declared that the workers and the drones should each build a comb. He would then decide who owned the original comb by comparing the cells of the comb and the taste of the honey in each case. The workers readily agreed, but the drones refused because it was beyond their capabilities. Judgement was immediately given in favour of the worker bees!

In medieval times, it was not just the honey that was important. Beeswax was needed to provide candles for churches and other religious establishments, not to mention the homes of the wealthy. The poor could not afford it; they used tallow candles made from animal fat.

Temporary hives or skeps were used by early beekeepers. Some of these were made of plaited straw, while others were constructed of hardened mud. On the next pages, there are photographs taken at rural museums, which illustrate the use of skeps. On Page 10, both types of skeps are shown, while Page 11 shows a traditional bee wall for housing the skeps.

It was difficult to extract the honey from skeps because it was hard to separate it from the bees. Some beekeepers, such as Charles Butler in the seventeenth century, made an effort to drive the bees out to an empty skep first. Most, however, held their skeps over burning sulphur to kill off the bees before they took the honey.

In the nineteenth century, various discoveries were made that were to revolutionise beekeeping practice, enabling honey to be extracted without harming the bee colonies, and also simplifying the various procedures that are involved with beekeeping.

9

Examples of old skeps at a rural museum.

In 1851, an American called Lorenzo Langstroth discovered that there is such a thing as a 'bee space'. In other words, if there is a space of 8mm, the bees will not build wax comb across it, nor will they fill it with propolis (bee gum). Interestingly, a smaller space will be filled with propolis, while a larger one will be covered with wax comb. This discovery meant that moveable frames could be used, allowing the frames to be lifted out of the hive, without harming the bees.

In 1857, sheets of foundation wax were developed by a Dutchman called Johannes Mehring. These were already imprinted with the hexagonal pattern that bees use when they build their combs, thus providing them with a ready-made basis or template for their constructions.

In 1865, a Frenchman, Abbé Collin, designed the first queen excluder. This was a thin sheet of perforated metal that allowed the worker bees to pass through from the brood combs to the honey combs above, but excluded the larger queen and the drones. This meant that the queen could only lay eggs in the brood combs.

In 1891, Edward Porter, an American, came up with the idea of a bee escape. This is a small contraption that allows the worker bees to go down from the honey comb to the brood combs, but not to return. This could be fitted before the honey combs on the top were removed, ensuring that as few bees as possible were on the honey combs when they were removed. Three decades earlier, the Austrian Franz von Hrushka, had come up with the idea of using centrifugal force to separate honey from the wax.

A traditional beewall where skeps were placed for weather protection, with a wattle hurdle placed in front. Taken at a rural museum in the West Country.

It was in 1945 that the German Karl von Frish solved the mystery of how worker bees find their way to and from the hive, bearing in mind that they often travel several miles in their food foraging activities. He demonstrated that the bees communicate by means of a 'bee dance' based on the relative distance to the feed source, the angle of the wax combs and the direction of the sun.

Research continues into the fascinating world of the honey bee, and every year, new information becomes available. We have come a long way since the old belief that, *"Bees are the roaming children of a dead cow"*. (Archelaus).

Bee Dances

Round dance

Circles one way then the other

Message: Nectar source near the hive - up to 80 metres. Go out and get it!

Wagtail dance

Tail wagging run indicates direction of nectar source and distance from hive. The slower the wagging, the farther the source

Tail wagging movement

Message: Nectar source is 60⁰ in relation to the sun from the hive.

Types of Bee in the Colony

Queen   Worker   Drone

Parts of a Worker Bee

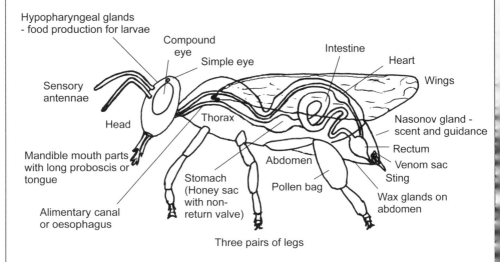

Hypopharyngeal glands
- food production for larvae

Compound eye

Simple eye

Intestine

Heart

Sensory antennae

Wings

Head

Thorax

Nasonov gland - scent and guidance

Mandible mouth parts with long proboscis or tongue

Rectum

Abdomen

Venom sac

Sting

Alimentary canal or oesophagus

Stomach (Honey sac with non-return valve)

Pollen bag

Wax glands on abdomen

Three pairs of legs

Structure of the Foot

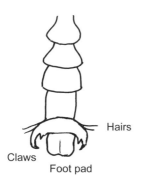

Hairs

Claws

Foot pad

Structure of the Mouth Parts

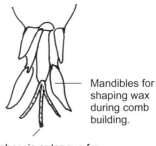

Mandibles for shaping wax during comb building.

Proboscis or tongue for sucking up nectar from flowers

Classification of the Honey Bee

There are many different kinds of bee, but the ones that we are concerned with are honey bees. We can see their relationship to the general classification of living organisms in the following table:

> *Kingdom*: Animal (as distinct from plants, viruses and bacteria)
> *Phylum*: Arthropoda (crustaceans, spiders and insects)
> *Class*: Insecta (all insects)
> *Order*: Hymenoptera (ants, wasps and bees)
> *Super Family*: Apoidea (all bees)
> *Family*: Apidae (long-tongued bees)
> *Genus*: Apis (honey bees)
> *Species*: Apis mellifera (European honey bee)

The Honey bee, *Apis mellifera* is the one that concerns us, but there are several races of this species, depending on the place of origin. The following are those that are appropriate to the northern hemisphere.

Black *Apis mellifera mellifera*

There are several variations of the Black European bee. The hardy native Black bee of Britain was virtually wiped out by Acarine disease at the time of World War 1 and European Black and Italian bees were brought in to build up the country's bee stocks again. The Black is a good forager but is nervous and easily disturbed, sometimes becoming aggressive.

Italian *Apis mellifera ligustica*

The Italian is lighter in colour and, although not as hardy as the Black, has the reputation of being more docile. Its foraging distance is claimed to be shorter than that of the Black bee.

Carniolan *Apis mellifera carnica*

The Carniolan is closely related to the Italian, but is slightly larger and darker. It has a good reputation for docility but is more prone to swarming than other races.

Caucasian *Apis mellifera caucasia*

The Caucasian also has a reputation for gentleness, as well as producing a relatively high level of propolis (bee gum).

These comments are generalisations for individual strains show differing characteristics. There has been a considerable amount of inter-breeding so it is true to say that most bees in Britain are of mixed races. New strains are always being developed, and many of the starter colonies that are available from breeders are hybrids based on several types, to include the best characteristics of each strain. Current research is based on developing Varroa-resistant bees.

Anatomy of the Honey Bee

Like all insects, bees have bodies that are divided into three sections, the head, thorax and abdomen. There are six legs (unlike arachnids such as spiders, that have eight legs). The different parts of the body have specialised functions, as follows:

Head

There are two compound eyes for overall vision and detection of colour, as well as three simple eyes (ocelli). These are thought to act rather like light intensity compasses in flight. The two feelers or antennae are for touch, taste and smell.

On top of the head are the hypopharyngeal glands that the worker bee uses to produce food for the larvae in the hive.

The mouth parts are equipped with a long tongue (proboscis) that is used to draw up nectar from flowers, while the mandibles on either side are used to shape the wax in comb building.

Thorax

This is the central part of the body to which the wings and legs are joined. The whole of the bee's body is covered with hairs. When a worker bee visits a flower, the pollen adheres very readily as a result.

Both the thorax and abdomen have openings called spiracles which lead into tubes called trachea. These are where air is taken in so that oxygen can circulate around the body during respiration.

Abdomen

The abdomen is made up of segments and is the largest part of the body. In the case of the worker bee there are four pairs of wax glands on the underside for the production of the wax needed for comb building and repair. On the top of the abdomen and near the end is a gland called the Nasonov gland. This produces a scent that other bees can follow and be guided to a nectar source. The fanning of the wings is also associated with this guidance.

At the end of the abdomen is the sting which is found only in female bees. The workers have a barbed sting which can only be used once because it causes the death of the bee. Once the sting has gone into the skin of the one being attacked, the barbs prevent its withdrawal without causing fatal damage to the surrounding area of the abdomen. (For further details see page 90). The queen bee has an unbarbed sting which can be used several times, but she uses it only on other queens. It is only the workers who are responsible for defending the hive.

Wings

There are two pairs of wings with the hind pair being smaller than those in front. In flight they combine to provide one surface and they are folded away when the bee is at rest.

As well as providing the means for flight, the wings are also used for fanning. As referred to above, this is associated with guiding other workers to a nectar source. The workers also fan their wings inside the hive, or at the entrance, in order to cool it in hot weather. They also evaporate water from collected nectar in this way, in order to produce honey.

Legs

The three pairs of segmented legs have feet that are equipped with claws and pads for walking on and adhering to various surfaces. The back legs of worker bees have pollen bags, an adaptation for carrying the loads back to the hive. The hairs on the legs act rather like brushes in transferring the pollen from the body into the bags. The front legs are adapted for combing and cleaning the antennae.

Digestive system

Food is taken in at the mouth and passes along the oesophagus or alimentary canal to the stomach where digestive juices and enzymes such as invertase break it down into simpler constituents. From here it passes into the intestine where digestion continues, and the necessary nutrients are absorbed into the body to cater for its metabolic needs. Waste products pass into the rectum from which they are periodically egested.

There is a specialised no-return valve between the stomach and the intestine. This allows food destined for the bee itself to pass through, but if nectar is to be carried back to the hive, it can be stored here without passing into the intestine. It is acted upon by enzymes and has some of the water absorbed from it before it is regurgitated into the wax cells.

Types of bee

Within the hive there are three types of bee: queen, drones and workers.

Queen

The queen's role is a purely reproductive one. She lays the eggs that produce the replacements for the colony. She is the biggest bee in the hive and has a longer, more slender body. Most of the time she has her wings folded over her back. She has no adaptations for collecting nectar and pollen and is fed by the workers. She does have a sting, although this is unbarbed and so can be used several times. It is only used against rival queens.

The Different Types of Cell in a Brood Comb

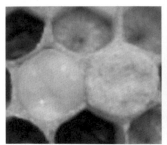 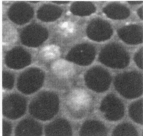 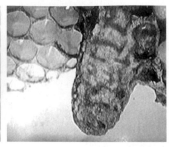

Worker cells are the smallest cells in a brood comb. The ones at the top are not yet capped.

The two capped drone cells at the front are bigger and more domed than the worker cells behind.

The queen cell is the largest and projects at an angle from the comb.

(Illustrations not to scale)

Drones

These are males that are only reared in the summer. They are easily recognised by their big eyes, an adaptation to spot the queen on her mating flight. They are bigger than the workers, are broader than the queen, and have blunt abdomens. They have no sting or any adaptation for collecting nectar and pollen. Their function is purely to mate with the queen. After mating they die. Any that are in the hive in autumn are pushed out.

Workers

The workers are under-developed females. As the name indicates, they do all the work. This includes comb building and maintenance, feeding the hive occupants, foraging for food and water, and guarding and cleaning the hive.

How the different bees are produced

All the bees in the hive are produced from eggs laid by the queen. So how is it that there are three different types of bee? It is all to do with the type of egg, the feeding and nurture. The queen gauges the size of the cell and lays eggs accordingly. She has the ability to withhold sperms in the sperm sac where they are stored, and in the larger cells she lays unfertilised eggs that will produce drones. Worker bees and queens are the product of fertilised eggs.

Queen Normal fertilised eggs are laid in specially built queen cells towards the edges of the brood comb. The cell in which the queen develops is extended to a length of around 25mm. It points downwards at an angle from the comb and looks rather like a pear-shaped peanut shell.

The egg hatches after three days and the young queen larva is fed on royal jelly (sometimes referred to as 'bee milk'). This is produced by the hypopharyngeal glands of the workers. On the eighth day, the cell is capped

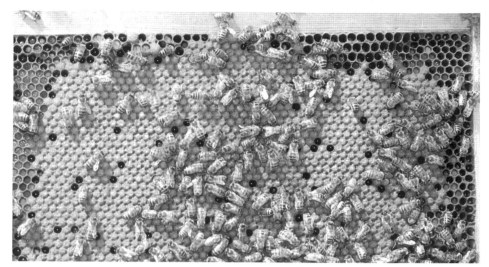
A brood frame showing capped and uncapped brood cells as well as stored pollen.

with wax.

Inside the capped cell, the larva become a pupa, a resting stage where all the activities are directed towards the development and eventual metamorphosis into an adult queen bee.

About a week later she emerges from the cell and is groomed and fed by workers. Her scent contains pheromones, sometimes referred to as 'queen substance'. The effect of these hormonal chemicals is passed to the other bees, communicating information that controls the orderly activities of the hive. After three days her wings are fully opened and she is capable of flight. She may use her sting to kill any other developing or newly hatched queen that she finds in the colony.

The virgin queen leaves the hive on her nuptial flight, pursued by the drones and mating usually takes place with several drones, on the wing. There may be several such mating flights. Enough sperms are stored in the sperm sac in her body to fertilise eggs for several years. After a time, she begins to lay eggs in the cleaned cells of the brood comb that the workers have prepared for her.

Drones Drones develop from unfertilised eggs, a process known as parthogenesis or virgin birth. As referred to earlier, the queen bee is able to withhold sperms from these eggs which are laid in hexagonal wax cells measuring 7mm in diameter. For the first couple of days the larvae are fed on royal jelly then on nectar and pollen for the next 10 days when the cell is capped by the

Stages in Development of the Honey Bee

 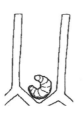 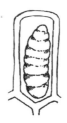 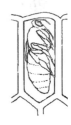

| After checking the size of the cell, the queen lays the appropriate egg. | The egg develops into a grub which is fed by the worker bees | The grown larva forms a pupa and the cell is capped | Development is complete and the bee bites its way out of the cell |

workers. The capping is convex shaped which makes it possible to distinguish drone cells from worker cells that are slightly flatter.

The drones pupate while they are in the capped cells, then 24 days after being laid as eggs, they emerge as bees. They are fed by the workers until they are ready to go out on the nuptial or mating flight with a new queen.

Workers They are first laid as fertilised eggs in hexagonal worker cells in the brood comb. Worker cells are around 5mm in diameter. When the larvae hatch, they are fed on royal jelly for a couple of days only. After two days, they are fed a mixture of pollen and nectar for a week until the workers seal the individual cells.

Inside the cells they pupate until ready to emerge which is three weeks after the eggs were first laid. Once the workers emerge from the cells their duties are organised as follows, depending on their age:

Day 1-2	Clean the vacated cells and help to keep the brood warm.
Day 3-5	Feed the older larvae.
Day 6-11	Feed the younger larvae.
Day 12-17	Make wax and build combs. Clear the hive of debris.
Day 18-21	Guard the entrance to the hive.
Day 22 on	Fly from the hive to collect nectar, pollen, propolis and water.

The queen bee has a much longer body than the worker and drone bees.

Worker bee with laden pollen bags on the legs.

Life expectancy

The average worker bee lives for around 6 weeks while it is foraging and working outside. In winter, in the hive, it lives for up to six months.

The drone lives for one season only. It dies after mating with the queen on her nuptial flight or, if it is still in the hive in the autumn, is driven out by the workers.

The average life expectancy of the queen is 2-3 years, but she may be replaced before this by the bees or the beekeeper if she is not laying properly. (See page 61 for further information).

Bee foods and raw materials

In order for the worker bees to be able to provide for the needs of the hive and its occupants, they need to find the necessary raw materials. They include the following:

Nectar

Nectar is a sweet fluid produced by the nectary organs of plants in order to attract insects for pollination purposes. It contains glucose, sucrose and fructose sugars. The nectaries of most flowers are found at the bottom of the petals. In some they may be in a funnel shaped corolla of petals so that the bees have to push their way down into the flower. Nectar is collected by the worker bees from whatever sources are available at the time of year. (See the *Bee Garden* chapter). It is sucked up through the bee's proboscis and stored in the stomach or honey sac. This has a non-returnable valve to ensure that it remains stored there while it is acted upon by the enzyme invertase produced by the bee's hypopharyngeal glands. Here the process of conversion into honey begins and a proportion of the water is absorbed.

The bee regurgitates the mixture into wax cells but before it becomes true honey it has to lose yet more of its water content by evaporation. The workers fanning their wings help to bring this about. Once ready, or ripe, the honey cells are capped with wax.

Honeydew

Honeydew, the sweet secretion of aphids, is also an important source of food, particularly in urban areas. The plane trees in London, for example, are especially attractive to the bees because of the aphids they shelter.

Pollen

Pollen is the main source of food for the bees. It is high in proteins, vitamins and minerals and enables all the processes of tissue growth and maintenance to take place. It is stored in wax cells by the workers. Their secretions together with a small amount of honey, preserves it in the cells until needed. In this state it is sometimes referred to as 'bee bread'. It is an important ingredient for the creamy brood food that the workers produce in their hypopharyngeal glands for feeding the larvae.

In northerly areas of Britain, in a particularly bad season where flowering may be delayed or the bees prevented from going out by inclement weather, it may be necessary to feed pollen, or a pollen substitute, to the bees in spring. (See page 54 for further information).

Water

A source of water is essential for the bees. The workers collect it and carry it back to the hive. Here, it is used to dilute honey so that the larvae can be fed. It is also required to cool the hive by evaporation in periods of hot weather. The bees will usually find a source of water, but it may be some distance away or it may be a pond where it is easy for them to drown. It could also be a neighbour's dripping tap with the resultant complaints that would bring! One solution is to provide an easily accessible source of water, not too far from the hive, and in a sunny spot. In shade, their body temperature soon drops and they may be unable to fly off again. A shallow container with stones placed in the water so that there is no danger of drowning, is suitable. Alternatively, a few pieces of floating bark or a piece of material draped from the water over the edge will provide safe access.

It is important to provide a water source for the bees early in the season. Once they have found a source, they will continue to use it, even though it may not be the best available. Old habits die hard.

A source of water provided in a sunny spot near the hives.

Foraging amongst the flowers.

Propolis

This is a brownish resin that is found as a protection on the scales of plant buds. (The horse chestnut 'sticky bud' is a good example). The workers collect it for use as a glue or varnish, hence the name 'bee glue'. It is carried to the hive as small globules on the bee's pollen baskets.

From the human point of view, propolis has medicinal properties and is also used as a varnish by violin makers. If necessary, it can be collected by placing a fibreglass mesh screen under the crown board. The bees will then cover the mesh with propolis after which it is removed and placed in the freezer for a few hours. When the mesh is removed and screwed up the propolis falls off.

The 'bee space,' referred to on page 10, is an all-important 8mm width which the bees will not fill with propolis, hence the ability to lift frames out of the hive for inspection and management.

Wax

Beeswax is the essential building material of the worker bees. With it they construct and repair brood cells and storage cells on the frames. To produce it, the workers need to eat honey in addition to their normal diet of pollen. Wax is then secreted from glands on the underside of the abdomen and the mandible mouth parts shape it as needed.

After honeycombs have been uncapped to extract the honey, it is a good idea to reclaim the wax for use again. Details are given on page 78.

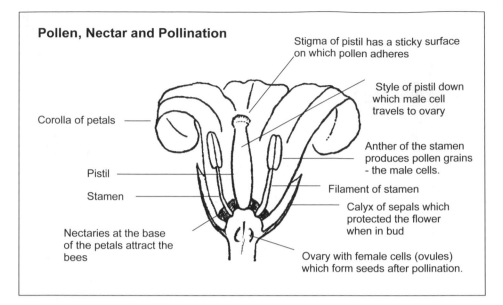

Pollen, Nectar and Pollination

Stigma of pistil has a sticky surface on which pollen adheres

Style of pistil down which male cell travels to ovary

Corolla of petals

Anther of the stamen produces pollen grains - the male cells.

Pistil

Filament of stamen

Stamen

Calyx of sepals which protected the flower when in bud

Nectaries at the base of the petals attract the bees

Ovary with female cells (ovules) which form seeds after pollination.

Pollination

Pollination is the process of transferring pollen (the male cells of a plant) from one flower to another so that fertilisation of the female ovules can take place. Bees are unwitting pollinating agents attracted to the flowers by the presence of nectar, pollen and brightly coloured petals.

As the bee pushes its way into a flower it brushes against the anthers or top part of the flower structures known as stamens. Here the pollen grains are produced and they collect on the bee's body hairs. When the bee subsequently goes to another flower some of the pollen is transferred to the sticky surface of the stigma which is the top part of the central pistil in the flower. From here, male cells can travel down to the ovary and fuse with the female ovules. After this fertilisation, the ovary with its seeds develops into a seed-head or fruit, as the case may be.

The honey flow

The term 'honey flow' is used to describe the period when the availability of nectar is at its peak, so it would be more accurate to call it the 'nectar flow'. It used to be the case that White Clover and Lime trees from June to July were the main plants involved, but widespread planting of agricultural Oilseed Rape in recent years has changed matters. Now, the main flow in these areas is in May, so beekeepers need to build up their bee stocks earlier than would traditionally have been the case in order to take advantage of it.

Foraging workers and guard bees at the entrance to the hive.

The *Bee Garden* chapter details many of the plants that provide bees with nectar, but here it is useful to look at the most important ones that make vital contributions to the spring, summer and early autumn nectar flows.

Spring flow	**March**	Willow	In recent years the main flow has become earlier.
	April	Fruit trees, Dandelion	
	May	Oilseed Rape, FieldBeans . .	
Summer flow	**June**	White Clover, Lime	Traditionally the main honey flow.
	July	White Clover, Lime	
	August	Heather	
Early Autumn flow	**September**	Heather	*Important sources not listed in this table include fruit plants, garden and wild flowers.*
	October	Ivy	

Types of Hive

National hive

WBC hive

Langstroth hive

Dadant hive

Smith hive

Photographs by courtesy of
E. H. Thorne (Beehives) Ltd

The Hive

Each cell in the honey comb has six angles,
as many angles as the bee has feet.
(Varro. Rerum Rusticanum. 36 BC)

We have come a long way since the days of bee colonies in hollow trees and skeps. Our apiaries are composed of a collection of purpose-made hives. They provide a protected environment for the bee colonies, just as the nests of wild bees did, but they are also designed to facilitate the handling and management of the bees by the beekeeper.

The key features of a modern hive are that it has moveable frames which can be fitted with sheets of wax foundation, there is a bee space of 8mm between the frames and the hive wall, and it is equipped with a queen excluder to separate the brood chamber from the supers above it. The brood chamber is where eggs are laid, while the supers are where honey is stored. There is also a roof with a crown board under it which prevents heat loss, and a floor with a bee entrance and entrance block.

The wood used in hive construction varies depending on the cost. Western red cedar is the most expensive but it does not need to be treated against damp. Softwood is cheaper but needs to be treated with an appropriate proofer. Clear *Cuprinol* that does not contain an insecticide is satisfactory, but do make sure that it is the insecticide-free one, for there are different versions.

Types of hive

There are several types of hive available and it is important to ensure that if frames are bought separately, they are the right ones for the particular hive. (Further details of a hive are shown on page 26, and details of frames are on page 34).

WBC hive

Often painted white, this is the type of traditional hive that most people visualise when they think of bees in a country garden. Although it looks attractive, it must be said that most beekeepers, these days, have the more modern and practical National hives.

The WBC hive, named after its designer William Broughton Carr, is still available and one of its characteristics is that it is double-walled. The outer wall is made up of a series of interlocking wooden 'lifts', while inside are the bee boxes. The double walls give the WBC extra weather protection which may be an important factor in some areas.

It is important to remember that while it may be warmer in winter it may also be cooler in summer. This could lead to the bees emerging later than

Parts of the Hive

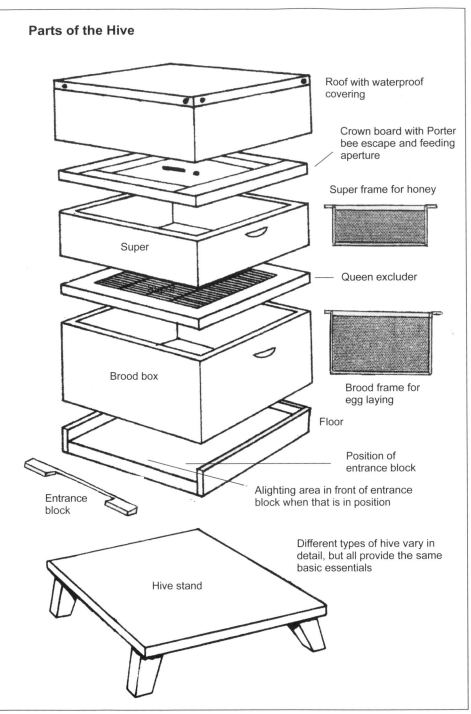

Roof with waterproof covering

Crown board with Porter bee escape and feeding aperture

Super frame for honey

Super

Queen excluder

Brood box

Brood frame for egg laying

Floor

Position of entrance block

Alighting area in front of entrance block when that is in position

Entrance block

Different types of hive vary in detail, but all provide the same basic essentials

Hive stand

would normally be the case. Because of this, the hive needs to be sited in a south-facing, sunny position. The WBC is heavy and unwieldy to manage, although it is equipped with feet which keep it clear of the ground. It is not practical if you plan to move hives on a regular basis.

National hive
This is the most widely used hive in Britain, and while it does not have the double-walled protection of the WBC, is much easier to lift, move and manage. It has convenient handholds at the sides of the box units and it uses British Standard frames which also fit the WBC.

It is possible to buy a hive stand and a gabled roof as extras from some suppliers, so that it looks as attractive as the WBC, but with the practicality of the National. It takes eleven brood frames compared with the WBC's ten, so there is more room for breeding.

Commercial hive
Sometimes called the Modified Commercial, this is similar to a National but is equipped with a deeper brood chamber for extra breeding space.

Langstroth hive
As far as the rest of the world is concerned, the Langstroth is the most popular hive. It is often used by commercial beekeepers, especially as it is possible to buy a Jumbo brood box, providing extra breeding space.

Dadant hive
Sometimes called the Modified Dadant, this hive is similar to the Langstroth but has deeper and wider frames, providing a large brood area. It is the largest hive available and is more suitable for the commercial beekeeper because it is very heavy.

Smith hive
Designed by Mr W Smith of Peebles, this hive is particularly popular in Scotland. It is in effect a miniature adaptation of the Langstroth and, being the smallest hive, is easy to move about. This makes it popular where hives are frequently moved to take advantage of the heather or orchard crops.

The Smith can take eleven British Standard frames but these are the ones with short lugs (supporting side bars at the top of the frame). Existing British Standard frames can be used if the lugs are shortened).

There are other types of hives available, but the ones that are most widely used are those indicated above.

This hive has been placed on a stand made of bricks and planks. An alighting board has been provided for the bees and an entrance block to reduce the entrance size is being put in position.

Details of the hive

Although hives are essentially just boxes to provide a safe environment for the bees, they have several sections that cater for different activities. Let us start at the bottom and work our way up.

Hive stand

Some hives, such as the WBC, come complete with legs. Most other hives need to be provided with a stand to keep them clear of the ground, away from rising damp and less accessible to intruders such as mice. Breeze blocks with wooden planks on top make perfectly good supports. It is also possible to buy purpose-made stands from suppliers.

Hive floor

This is a board on which the bottom section or brood box of the hive sits. It has an entrance through which bees enter, and an entrance block which can be used to reduce or close off the entrance. In winter it is advisable to add a mouse guard to stop mice over-wintering in the hive. Some hives also have an alighting board projecting outwards from the front of the floor.

Entrance block in position. In winter a mouse guard is also used to keep the small rodents out.

Open mesh floors (OMFs) are also available for National hives. These have a plastic tray that slides under the galvanised wire mesh. The advantage of them is that they help to monitor and control the Varroa mite which is parasitic on bees. The mites fall through the screen but cannot climb back up again. The number on the tray is an indicator of the level of infestation. For further details of the Varroa mite see page 70.

OMFs are also said to improve ventilation by allowing water vapour to escape through the floor.

Brood box

Placed on the floor is a brood box, sometimes referred to as the brood body or brood chamber. This has a series of vertically hanging frames on which the queen bee lays her eggs. Here may be found workers going about their various tasks of feeding and looking after the larvae, queen and drones. Further details of the frames are given below.

Queen excluder

Placed on top of the brood box is a queen excluder. Made of wire, slotted steel or plastic, this stops the queen from going up into the super above and laying her eggs there. Drones are also too big to go through. Only the smaller workers can pass through the apertures so that they are able to make honey and store it in the frames of the super. An excluder is not required in winter but should be in place by mid-April.

Roof

Position of Crown board

Super

Super

Queen excluder

Brood box

Entrance

Alighting board

This hive has two supers above the brood box, providing extra space for storing honey. In a good year, even more supers may be required.

A queen excluder which allows worker bees to go through to the supers above the brood box but which prevents the queen and drones from doing so.

Super

On top of the queen excluder is a box called the super (from the Latin meaning above). It also has vertically hanging frames but these are shallower than those of the brood box. It is here that the worker bees make and store most of the honey. Several supers can be placed one on top of the other, allowing extra space for the honey, if necessary. In a good year, three or four supers may be needed.

Crown board

Placed on top of the supers is a crown board. This provides insulation, to stop too much warmth escaping upwards. An aperture on the board allows a feeder to be placed on top when supplementary feeding is required.

The crown board can also be fitted with one or more Porter bee escapes which allow worker bees to go out but not to come back. This is useful when the honey crop is to be taken and it is advisable to have as few bees on the supers as possible.

Crown board being placed on top of the supers.

Roof

Finally, a roof is placed over the crown board. This is waterproof so that the hive is protected against the vagaries of the weather. There are ventilation holes at the sides, with metal gauze to prevent insects such as wasps and robber bees getting in.

Frames

There are two kinds of frame: those that are designed to fit into a brood box where eggs are laid by the queen, and those that go into a super where the honey is stored. Brood frames are deeper than honey frames to allow for the maximum amount of breeding.

The roof above the crown board has a waterproof covering on top, as well as a ventilation panel.

The ventilation panel in the roof allows air in and out but excludes insects

Each frame hangs vertically by its lugs (supporting end bars) within its box and has a bee space of 8mm at each end. This is the critical space referred to earlier which ensures that a frame can be lifted out of the box. A smaller space would be filled with propolis gum by the bees, while a larger one would have wax comb built in it. The frames also need to have a bee space between each other. Metal end spacers which fit over the lugs have traditionally been used, but plastic spacers are now available. There are also Hoffman self-spacing frames which, as the name indicates, do not need end spacers. They are usually found as standard in the Modified Dadant, Langstroth and Modified Commercial hives, or as options for the WBC, National and Smith hives.

33

Frames

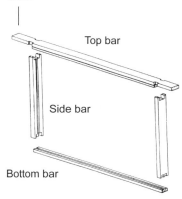

Lugs at either side of top bar for suspending the frame

Top bar

Side bar

Bottom bar

There are two kinds of frame in the hive: the brood frame for the brood box and the super or honey frame for the super.

They are suspended vertically and parallel with each other in the appropriate box.

British Standard Frames

Metal end frame

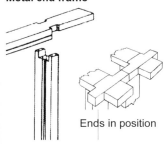

Ends in position

Metal or plastic ends are needed with this type to provide the correct spacing between the frames.

Manley frame

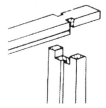

The side bar is wider so that the frames are self-spacing when in use. (Super only)

Hoffman frame

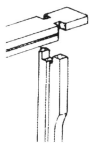

The Hoffman is also self-spacing.

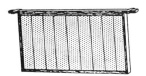

Frames are available already assembled with foundation wax or you can assemble them yourself using gimp pin nails.

Foundation wax reinforced with wire, ready for inserting in a frame.

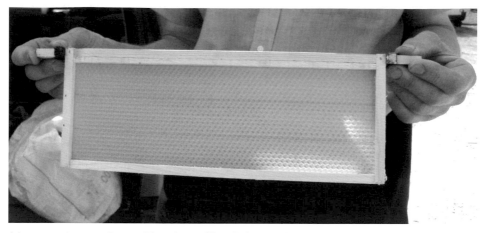

A home-made super frame with a sheet of foundation wax inserted, and with metal ends placed on the lugs at each end of the top bar.

Assembling a Frame

| Remove wedge from top bar. Assemble frame with one bottom bar. | Nail with 2 nails for side bars. Nail bottom bar up through side bar when using Hoffman frames. | Bend over long loops of foundation and lay foundation onto top bar | Nail wedge onto top bar trapping the bent wires. Nail through each loop of wire and the wedge at an angle. Nail in second bottom bar. |

Illustrations by courtesy of E H Thornes (Beehives) Ltd

Reference has been made to the fact that if frames are bought separately, they should be ones that fit. If you have doubts about this, experienced beekeepers and suppliers will advise.

Frames can be bought already assembled and complete with foundation wax, ready to go into the hive. Cheaper alternatives are to buy them with parts for you to assemble yourself, as shown above. Those who are interested in DIY may even want to make their own wooden frames, such as the one in the photograph at the top of the page, but there is not a great saving in doing this.

Protective clothing and equipment

Protective clothing really is essential, especially for the beginner who has not built up any resistance to bee stings. We have all seen photographs of old beekeepers who open up hives without wearing gloves and a veil, but I have always regarded this as bravado. Even the most docile of bees can occasionally be bad-tempered. Essential items of clothing are:

Bee suit

Purpose-made suits are usually made of close fabric polyester cotton that is white in colour. White is inoffensive to bees so multi-coloured outfits are best avoided. The material is smooth, not only to keep bees out, but also to prevent them becoming entangled in the fabric. The wrists and ankles of the suits are often elasticated for greater security. Some are pull-over designs while others are zipped in the front.

If a bee suit is not available, ordinary clothing can be worn as long as the materials are not wool, are heavy duty, fasten securely and have long sleeves.

Boots

Rubber boots are ideal, but ordinary boots can be used as long as they allow the trouser legs to be tucked in securely. Bees crawl upwards so if they gain access to a trouser leg it can be painful.

Veil

There are several types of bee veil available, some which are designed to fit over a hat and all-in-one models that include the hat. Some have rings to keep the veil well away from the face and head. Key points are that they offer unrestricted vision and fit securely. This might be by means of zips, velcro or elasticated straps placed under the arms. If you have the opportunity to try some on before buying, so much the better.

Gloves

Gloves may be of leather, kid or plastochrome, depending on how much you are prepared to pay. Needless to say, the more expensive ones are less cumbersome than the cheaper ones. The old saying, 'fitting like a kid glove' is certainly appropriate here. Bee gloves have elasticated gauntlets that pull up to the elbow for added protection. Gloves should be comfortable and well-fitting otherwise it will be difficult to work with them.

Smoker

Smoke has a subduing effect on bees. In the wild, if a forest fire threatens a colony, the instinctive response is to feed in order to provide extra energy for escape and movement to a new site. Our domestic bees have the same instinct

Protective clothing is essential. These beekeepers are kitted out in bee suits, veils, boots and gloves with arm gauntlets.

and when well-fed they are less likely to sting. The minimum of smoke should be applied but from the beekeeper's point of view, a smoker that produces enough smoke during the whole process of hive inspection is essential. It would not do for it to go out at a crucial time. Cool smoke is required for if it is too hot, it will damage the bees and may even make them aggressive.

Smokers should be well alight and operational before the gloves and veil are donned for obvious safety reasons. Crumpled up newspaper can be used to light the smoker initially, with other fuels added to produce the smoke. Materials that can be used include: corrugated cardboard, dried grass, old non-synthetic rags, hessian or slivers of well-dried wood. Get some practice

Two types of smoker in common use. Both operate on a bellows principle that puffs out smoke where necessary but the blunt-nosed type is more efficient than the older type on the left.

in lighting and getting smoke effectively before putting it to use.

• Apply a few puffs at the hive entrance then wait a few minutes to give the bees a chance to respond.

• Puff smoke across the top of the frames when these are revealed.

• Be ready to apply more smoke as needed during the inspection.

• Use the minimum necessary. How much comes with experience.

• Some beekeepers like to use a very fine waterspray to subdue the bees, but it is important not to overdo it and soak them.

Hive tool

This is a flat piece of steel with a wide, flat blade at one end and a curved surface at the other. It is used to loosen the frames when they are being taken out of the hive and is effective for scraping away the propolis.

Bee brush

A very soft hand brush is useful for gently brushing bees off a frame without harming them. Purpose-made ones are available from bee suppliers or you can do what I do and use a quill feather.

Feeder

There are times when it is necessary to provide food for the bees. In spring, for example, there may be an inadequate stored supply after over-wintering. After removing the honey in summer and early autumn it is also necessary to provide a replacement for winter stores.

There are several different kinds of feeder available, including a drawer-type that covers the top of the brood box, but this is more appropriate for the

Smoker being used to pacify the bees when the hive is opened.

Hive tool being used to gently loosen a frame before it is taken out of the hive box.

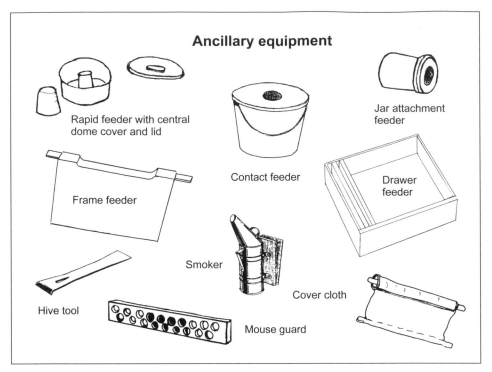

Ancillary equipment

Rapid feeder with central dome cover and lid

Jar attachment feeder

Frame feeder

Contact feeder

Drawer feeder

Smoker

Cover cloth

Hive tool

Mouse guard

commercial beekeeper. Then there is a frame feeder that hangs parallel with the brood frames. For the small-scale beekeeper a rapid feeder that sits on top of the feed hole in the crown board is fine. There are also contact feeders with lids. Once filled, these are inverted and placed on the feed aperture of the crown board. It is also possible to get a jar attachment feeder for feeding small quantities. This is designed for use with a glass honey jar and is suitable for a small nucleus of bees.

Cover cloths and wedges

Cover cloths are useful for covering the top of the brood box when the hive is opened for frame inspection. The cloth is placed over the frames after the queen excluder is removed and the first frame is removed. It helps to prevent loss of warmth if the day is a little chilly and also keeps the bees on the frames rather than flying upwards during inspection. Wedges are useful when opening up a hive and for all sorts of other tasks.

Cover cloths are easy to make from any smooth heavy duty fabric that rolls up easily. Sewing in a wooden rod at each end stabilises the cloth when it is in position so that a gust of wind does not blow it away.

Getting Started

Nine bean rows will I have, a hive for the honey bee,
And live alone in the bee-loud glade.
(W B Yeats. The Lake Isle of Innisfree. 1893)

The local society

If there is a local bee society then do join it! It may be a local branch of the county society which, in turn, may be associated with the national *British Beekeepers' Association* (BBKA). Here you will be in contact with others of the same interest, have access to help, advice and information, and be able to attend interesting talks, apiary visits and practical demonstrations. It is often possible to hire larger equipment such as honey extractors from local groups, as well as buy in supplies such as honey jars and labels. Membership of the BBKA also has the advantage of insurance cover for its members.

Acquiring hives

Although one hive will cater for the bees, it is best to have at least two, so that in the event of swarming, there is a place to house a new colony. There are three options when it comes to acquiring hives:

Buying a new hive

An easy approach for a beginner is to buy a complete starter package from a specialist supplier, although this will obviously be dearer than setting up a secondhand hive. A starter package will include the hive, frames and some bees. Hives are also available ready-assembled or as flat packs for you to put together yourself following the instructions from the manufacturer. Check that all the parts are there! The assembled hive will need to be treated with an insecticide-free proofer before use.

Buying a second-hand hive

Local societies are the best place to find out if there are any second-hand hives in the area. The *Classified Advertising* section of local newspapers may prove to be useful, and there are also national beekeeping and smallholding magazines which have *For Sale* advertisements.

It is probably possible to buy second-hand hives on website auction sites, but great care is needed, not only to establish their condition, but also to ensure that there are adequate safeguards for credit card purchases on-line. A hive should be free of damp, rot and other conditions which may have had a degenerative effect on the wood.

A dummy board is a plain board set in a frame which can be hung in a brood box to confine a 'nuc' to one area of the chamber. Plywood or even cardboard can be used in the frame if making your own.

A 'nuc' consists of a few frames of brood comb with a queen, attendant workers, as well as eggs, larvae and sealed brood.

Making your own hive

Those with DIY skills may want to have a go at making their own hives. If so, there are some excellent plans available from the *British Beekeepers' Association* for doing so.

Acquiring bees

The bees may be supplied as part of a starter package or they can be bought separately. They are available in one of the following ways:

Nucleus

Usually called a 'nuc' this is the best way for a beginner to acquire bees. It is often available as part of a starter kit. It consists of about four frames of brood comb with a young mated queen and attendant worker bees. There should also be some eggs, larvae and sealed brood with adequate stored food on the frames.

When delivered in spring, the frames will be in a travelling box with instructions on how to remove the frames and place them in the hive. They are hung in the brood box, but as there are not enough of them to fill the chamber, it is a good idea to use a dummy board. This is a plain board frame which confines the bees to one area, ensuring that they are able to control the hive temperature without difficulty. As the colony increases, an empty foundation frame can be inserted in place of the dummy board, while that is moved along. In this way, the colony is increasing in size at the same time that the beekeeper is acquiring confidence and experience.

Commercial suppliers of nucs will have treated them against Varroa using methods approved by DEFRA.

Colony

This is a hive with an existing and full colony. More often than not it will be from a local beekeeper who is retiring or cutting back on his activities.

Joining the local bee society is an excellent place to start for the beginner. Here members of the Saffron Walden branch of the Essex Beekeepers' Association are attending a practical demonstration at a local apiary.

Needless to say, the hive will need to have the entrance and crown board holes blocked for transportation. Purpose made hive straps and lock slides are available from suppliers which make it much more secure.

The hive will be heavy to move and you will need help to transport and set it up. Get assurances from the seller that precautions against Varroa have been taken and that the colony is healthy and active. The seller should also be prepared to help you with the various tasks such as hive inspection, feeding, swarm prevention, etc.

Starting with a swarm
It was mentioned earlier in the book that bees belong to you while they are on your property. If they should swarm and land somewhere else, they can be taken by anyone.

Starting beekeeping by trying to take a swarm is not recommended for the beginner. It is best left until more experience is gained. It could be, of course, that a member of the local beekeeping association takes one for you. Checks on the bees' health are advisable. For example, it is a good time to put in a Varroa strip. (See page 70). A sample of bees may also be taken for inspection by the DEFRA bee inspector for the area. Alternatively they can be sent to the National Bee Unit (NBU). (See Reference section). The way in which a swarm is taken is detailed on page 58.

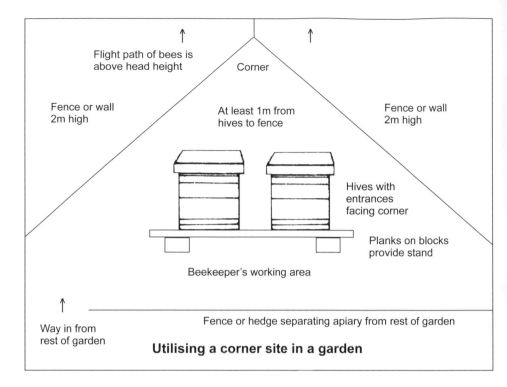

Utilising a corner site in a garden

Diagram labels:
- Flight path of bees is above head height
- Corner
- Fence or wall 2m high
- At least 1m from hives to fence
- Fence or wall 2m high
- Hives with entrances facing corner
- Planks on blocks provide stand
- Beekeeper's working area
- Way in from rest of garden
- Fence or hedge separating apiary from rest of garden

Siting a hive

Hives need to be sited in an area that is sheltered from strong winds but has sunlight for at least part of every day. Avoid areas that are known to be frost pockets. The flight path of the bees should be well above the heads of any neighbours or passers-by. Consideration towards neighbours is important. For example, try not to open hives when they are sunbathing or the children are playing in the garden. Above all, keep calm and gentle bees.

Having a tall hedge, fence, shed or greenhouse about a metre's distance from the front of the hives will ensure that the flight path is steep enough, but make sure that the entrance is not on the same side otherwise you will have to dodge the bees yourself. In a garden, utilising a corner, with the hive entrances facing the corner is appropriate. Wattle fences provide good screens for hives. In fact, experience has shown that if neighbours are not able to see your hives they are less likely to worry about them. Avoid privet hedging for it can give a bitter taste to the honey.

Many beekeepers have out-apiaries, although they tend to be experienced or commercial beekeepers. Out-apiaries are where hives are placed outside the property: in an orchard, on farmland for oilseed rape, or on moors for heather. Security is an issue here, for beehives have been stolen.

An out-apiary also needs to be accessible by car otherwise moving the hives can become difficult or even impossible at certain times of the year. Further details of out-apiaries and moving hives are on page 65.

Setting up the hive

To minimise backache from bending, the hives need to be about 30cm (1ft) off the ground and placed on a stand, either purpose-made or planks on blocks. This not only protects the hive against rising damp but also provides weed-free access to the alighting area. If grass or weeds do grow tall, cut them back. At the back of the hives there should be sufficient room for the beekeeper to gain access and work with them comfortably. A dry garden shed is ideal for storing bee equipment.

Make yourself ready with protective clothing and light the smoker before putting on the veil. Place the floor and brood box on the stand. If a nuc has been purchased, remove the frames and insert them in the centre of the brood box, with an empty foundation frame on either side. A dummy board can be placed on either side in order to confine the bees to this central area initially. As the queen lays more eggs and the colony expands, the dummy boards can be removed and replaced with empty foundation combs.

Check that the queen is present and that there is brood and food, as well as some attendant worker bees. (The following pages indicate how this is done). Feeding at this stage is important. The chapter on *Feeding* gives instructions on how this is carried out. Check for any evidence of disease, as indicated in the chapter on *Pests and Diseases*. Leave the bees alone for at least two weeks to settle in, before opening the hive again.

A queen excluder and super need only be placed on the brood box once all the combs in the brood box have been drawn. When starting with a nuc, the emphasis needs to be on building up a strong colony. The honey crop will obviously not be great in the first year.

Fanning the wings to provide ventilation.   Poached egg flowers are favourites in the garden.

Handling and Checking

If the hive be disturbed by rash and stupid hands,
Instead of honey it will yield us bees.

(Ralph Waldo Emerson. 19th Century)

It should be emphasised that a practical course is essential before embarking on beekeeping. This small book does not claim to be a comprehensive, practical primer. Rather it aims to provide a basic overview of the tasks involved and to point you in the right direction. The first step is to join the local bee association and apprentice yourself to an experienced beekeeper who will take you through tasks such as the ones listed below.

During the active period between late-March/April and October regular checks of the hive are advisable, not only to ensure that everything is as it should be, but also to prevent problems from developing.

• **Is the queen present?**
If there are eggs present, then so is the queen. If she can be found, so much the better. She can be marked in such a way that she is easy to spot in future. (See page 61 for details).

• **Is she laying plenty of eggs?**
If you have bought a nuc from a specialist supplier there will be eggs, larvae and sealed brood cells apparent on the frames.

If no brood is apparent, the queen is either dead or not laying. A new queen must be introduced. (See page 61).

If there are a large number of drone cells, the queen is probably running out of stored sperms and will need to be replaced.

• **Is the colony strong and productive?**
Not only should the queen be busy, with eggs, larvae and sealed brood apparent, but the workers should be flying in and out, foraging and coming back with laden pollen baskets. They will also be seen to be carrying out all the other maintenance tasks such as comb building and cleaning, feeding larvae, and so on.

• **Is there enough food for them?**
The workers should be storing pollen in some of the cells, as well as bringing in water and nectar for producing honey. They tend to store honey on the edges of the brood frames, especially the corners, while the eggs and larvae are more centrally placed. Pollen varies in colour depending on the plant source.

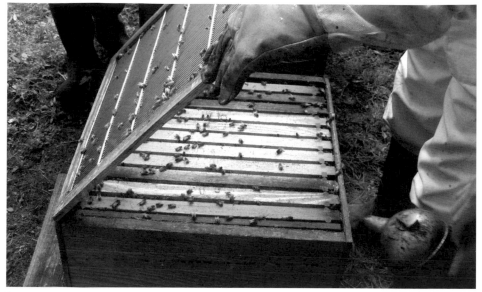
Be ready with the smoker!

Experienced beekeepers can usually judge the amount of stores by the weight of the hive. It is not necessary to lift up the whole structure but only one side of it, a process known as hefting. This is something that is best demonstrated by a member of the local beekeeping association.

• Is there any sign of disease?
Regular checks are needed, not only to see if there are Varroa mites present, but also for a range of problems that can appear. Details of what to look for and what needs to be done are given in the chapter on *Pests and Diseases*.

• Is there plenty of room for brood and food?
If you started with a four-frame 'nuc', it is necessary to add empty foundation frames on a regular basis, to cater for expansion of the colony. If it is an established colony, it is also a good idea, in my view, to put a super above the brood box as soon as possible in the year. I usually have mine in place by the middle of April. This gives the workers room to store their honey, without taking up valuable space on the brood frames.

• Are the bees unduly angry?
Bees become angry for several reasons. It could be the weather. The colony may have swarmed and they are without a queen. The nectar flow may be insufficient or there may be hive robbing going on. The queen could be old, but it will take several inspections to establish whether the colony is consistently bad tempered. If it is, then re-queening is usually the best option.

Stages in Opening and Checking a Hive

1. Choose a mild day (above 16°C) with no wind, and in the early afternoon when many of the bees will be outside.

2. Wear protective clothing and have hive tool and smoker ready before donning veil.

3. After applying smoke at the entrance lift off the roof and place it, upturned, towards the hive front.

4. If there is a super already in place remove it with the crown board on top. Place them on the upturned roof.

5. Lift off the queen excluder

6. Continue to apply smoke as needed.

7. Use the hive tool to gently free the frames. If there is a dummy board at the side, remove it to create access. If not, remove a frame and shake it free of bees at the front of the hive.

8. Lift out each frame holding it by its lugs and examine both sides. Replace it in the same position as before.

9. Examine the brood frame carefully.

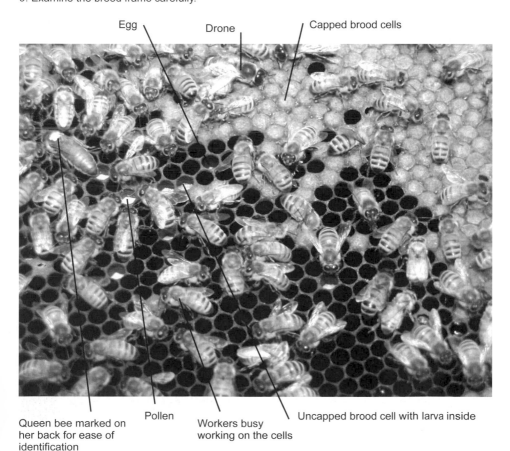

Egg

Drone

Capped brood cells

Queen bee marked on her back for ease of identification

Pollen

Workers busy working on the cells

Uncapped brood cell with larva inside

10. Try and identify all the different types of cells.

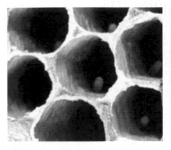

Cell with eggs

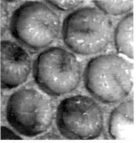

Larvae just before cells are capped

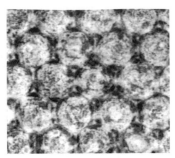

Capped brood cells

Brood pattern on a comb

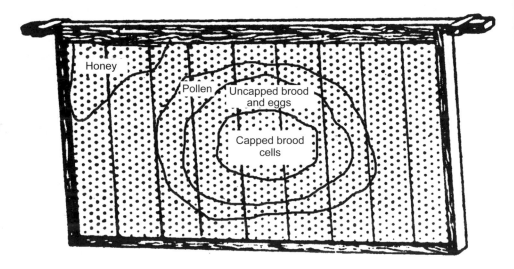

The diagram above shows the general pattern of cell usage on a brood comb, but it will vary over time. For example, as the bees emerge from the central capped cells these will be cleaned by the workers and made ready for the queen to lay more eggs. The area currently shown as uncapped brood will then have been capped, with the next lot of larvae emerging, and so on.

Honey tends to be stored in the corners of a brood frame, while pollen is often placed beyond the brood cells. However, this is a generalisation, for it may also be stored in any available empty cells. Pollen varies in colour, depending on the plant source, eg, yellow in dandelions, orange-red in horse chestnut and grey-blue in bluebells, so is fairly easy to identify.

Capped and uncapped honey cells

Honey cells are usually built in the corners of brood frames.

If you have any doubts at all about being able to read the combs and identify the various elements, ask an experienced beekeeper from the local association to show you. He or she will also be able to advise on what procedures to follow if there is a potential problem.

In the chapter on *Pests and Diseases* (Page 86) illustrations are provided to help you identify specific problems.

Allowing for expansion

If you bought a four-frame nuc, it is obviously important to make extra brood foundation frames available as they become necessary. The dummy board will eventually be removed entirely so that the whole brood box is available for the queen to lay eggs. In the spring, expansion can take place very rapidly so it pays to be observant.

Some hives can take deeper brood frames to cater for colony expansion, while some beekeepers use a super to provide extra room. In this case, the queen excluder would obviously be placed above that super and under the honey super. In most cases, however, and certainly in the case of the small-scale beekeeper, the normal brood chamber usually provides enough space.

Feeding

So the industrious bees do hourly strive
To bring their loads of honey to the hive.

(The Woman's Labour. 1739)

The main reason for keeping bees is obviously for the honey crop, but when this is taken away, the bees must be provided with an alternative source of stored food for the winter. There are other times when they need feeding, too. In poor seasons, or where spring is late and cold, providing food will not only help the bees through a difficult patch but it will also help to build up the colony. Feeding is best done in the evening as the bees become excitable when there is food around.

Food

The food provided for the bees needs to be in a form that is easily accessed and assimilated by them. The options are as follows:

Sugar syrup This is ordinary white granulated sugar dissolved in warm water that is then left to cool. Brown sugar should not be used because it can adversely affect their digestion. For spring or emergency feeding, the ratio for making up a solution is a one kilo bag of sugar to every litre of water. For winter feeding, the ratio is 1kg sugar to 750ml water.

Syrup is ideal for autumn and spring feeding. In autumn the brood cells are emptying and can be used by the bees for winter stores. In spring the syrup is required for the rapid increase in brood rearing

Fondant paste This is available from bee suppliers or bakers and is placed directly on the top bars of the brood frames. It is particularly useful for winter use. Syrup at this time may excite the bees so that they leave the hive and then die of cold.

Feeders

There are several different types of feeders available and it is a matter of personal preference as well as the scale of beekeeping operations that dictates the choice. They are illustrated on page 40 and include the following:

Contact feeder Available in different sizes, this is a lidded container that when filled with sugar solution is upended on top of the feed hole in the crown board. The lid has a circular gauzed area that allows the syrup to permeate slowly. As the roof cannot then be placed on top of the feeder, a super minus its frames can be used to support it. If you do not wish to use a super, it is a simple matter to make an eke. This is just a four-sided wooden box without a floor or a roof, made to fit the hive. When placed around the feeder it then supports the roof.

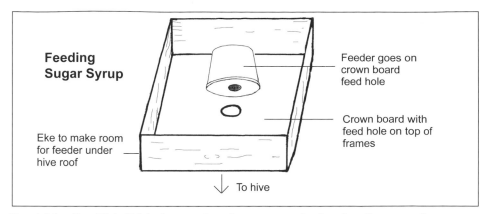

Feeding Sugar Syrup

Feeder goes on crown board feed hole

Crown board with feed hole on top of frames

Eke to make room for feeder under hive roof

To hive

Rapid feeder This lidded container has a central tube that fits over the crown board feed hole with an eke on top. When filled with syrup, the feeder allows the bees to crawl up the tube which has a cover over it to prevent drowning. The bees can access the syrup more readily than with a contact feeder. The advantage of this type of feeder is that it can be refilled by taking off the lid without having to remove the feeder from the crown board.

Frame feeder This is a hollow frame that is filled with sugar syrup then inserted in the brood box where it hangs vertically with the other frames. Wooden strips or plastic mesh prevent the bees drowning. It is not appropriate for cold weather where opening the brood box would chill the bees.

Jar attachment feeder This works like a contact feeder and is designed to take a standard glass honey jar. When upended onto the hole in the crown board it provides bee access to the syrup via the gauzed area. It is suitable only for feeding small quantities at a time.

Tray feeder This is a tray or drawer that sits on top of the brood box. Bee access is at one end of the feeder. It is appropriate for feeding large amounts of sugar syrup and is often the choice of commercial beekeepers. There are several different makes and it is important to get one that fits the hive.

Spring feeding

For an established colony there should have been sufficient stores to get the bees through the winter, as long as they were provided with food in the autumn. Experienced beekeepers can judge this from the weight of the hive, by hefting. To do this, place your hand under the floor at the back of the hive and lift it slightly so that it tilts a little way, just enough for you to sense the weight. If it feels light then feeding is required. The technique is best acquired by getting an experienced beekeeper to demonstrate it, and to be there while you try it out.

From the middle of March to mid-April is a crucial time, for the queen will be laying and the colony is expanding rapidly. If you bought a hive with a nuc, then obviously they will not have enough stored food and must be fed. Around 4.5 litres (1 gallon) of sugar syrup per hive is the average amount needed (A drawer-type feeder or a large contact feeder will take this in one go, but a smaller feeder will need to be kept topped up).

Before the middle of March it is best to feed fondant paste which can be placed directly on top of the brood frames, in case the bees are excited to the extent that they leave the hive in unsuitable weather conditions.

Hot water from the hot tap in the kitchen is suitable for making up the sugar solution. Use a one kilo bag of white, granulated sugar to every litre of water and stir it well, then leave to cool before use. The amount you make up at a time will obviously depend on the size of your feeder. To carry a large amount out to the hives, a 5-litre plastic water container from the supermarket is useful.

Autumn feeding

A strong colony of bees needs around 17kg of honey to see it through the winter. If the honey is removed, the deficit must be made up by feeding sugar syrup. This should be made available after the supers are off and the last honey has been harvested. The bees need enough time to store, ripen and seal it before the winter. Unsealed sugar syrup can absorb water and become mouldy, or even ferment in the cells. The ratio at this time is 1kg sugar to 750ml of water.

Feeders that can take at least 4.5 litres (1 gallon) of syrup at a time are preferable to smaller ones, because it is surprising how quickly the bees will take down the syrup. When it becomes apparent that no more is being taken, then you can stop feeding, and certainly by the end of October.

Pollen feeding

In most years there are enough flowers early in the year to provide pollen. In northern areas, or where the season is particularly poor, it may be necessary to feed pollen. Some beekeepers harvest a proportion of the pollen by using pollen traps. This is where a grille is used that the workers must pass through. As they do so, some of the pollen is scraped off their bodies and falls to a collecting floor underneath. This technique is usually adopted by commercial beekeepers who may be selling pollen for medicinal purposes. Where bees are short of pollen in the spring it is possible to feed a pollen substitute that is high in protein and vitamins. Available from suppliers, this is based on soya flour and when mixed with water to form a paste, can be put on top of the frames over the brood area.

Swarming

A swarm of bees in May is worth a load of hay,
A swarm of bees in June is worth a silver spoon,
A swarm of bees in July is not worth a fly.

(Traditional)

Swarming is an instinctive pattern of reproductive behaviour in order to form new colonies. Some of the eggs that the queen has laid will produce new queens. A swarm takes place when the old queen leaves the hive with about half the bees in the colony, leaving a new queen to take over in the hive. One of the most important of the beekeeper's tasks is to be prepared for swarming, take action to prevent it if possible or cope with it when it happens.

Be prepared!

From spring onwards drones will begin to appear in the hive. This is part of the spring routine and does not mean that swarming is imminent. Nor does it mean that the queen is short of sperms, as referred to earlier. It is only where there is an excessive amount of drone brood by comparison with other brood, that this may be the case. However, the activity may be regarded as the first sign of swarming to come, so checking the hive every week to ten days is a good idea at this stage.

Queen cells

A more immediate indication of swarming is when queen cells begin to appear. Initially, they look like acorn cups but as they are extended they resemble peanut shells. They tend to be built around the edges of the comb and hang vertically. However, it is important to bear in mind that queen cells are produced for a number of reasons. If the queen has been killed by accident, as for example by clumsy hive checking, emergency queen cells will appear very quickly. These are fairly easy to recognise because they are made in worker cells in the brood frame and protrude outwards from the frame before bending down vertically.

Where a queen is old or failing in some way, she produces less pheromone. The colony recognises this and produces queen cells as replacements. Once a new queen has emerged, been mated and is laying successfully, the old queen will be killed. An old queen is also more likely to produce a swarm than a young one, another reason why most beekeepers prefer to replace a queen every two years.

Swarm clustered on a branch. *(Anna Chambers)*

Hiving a swarm. *(Anna Chambers)*

Overcrowding

If there is insufficient room for the colony, they will swarm to find a new home. Providing new brood foundation combs will attract a proportion of bees to work there, reducing the congestion elsewhere. Lack of egg-laying space is also a factor to be taken into consideration. It is possible to add a shallow super on top of the brood box, with the queen excluder above that, so that it provides extra laying space. Providing a super above the queen excluder is also necessary to cater for the nectar being brought in, otherwise the bees have no choice but to store it all in the brood frames. I always have a super on by mid-April.

It is possible to remove some brood frames with attendant bees and place them in a new hive. In other words, treat them as a nuc, as detailed earlier. The space left in the original is then filled with new foundation brood frames. (See page 62 for further information on creating a nuc).

Genetic tendency

Although swarming is a natural phenomenon, some strains of bees have less of a tendency to swarm than others. Ask the breeders about this.

Bad weather

Sudden bad weather which confines the bees to the hive, after a period when they have been bringing in a lot of nectar, can also induce swarming, probably because of the perceived over-crowding. There's not a lot you can do about the weather, but providing extra space as mentioned above, will deal with the over-crowding.

Preventing swarming

To sum up, although swarming will inevitably take place at some point, the chances are reduced by the following procedures:

- Replace the queen every two years.
- Make sure that there is enough room in the brood box.
- Put a super on early in the season (April) to cater for incoming nectar.
- Choose bees with less of a genetic tendency to swarm.
- Remove unwanted queen cells.
- Make an artificial swarm. (See page 60).
- Inspect the colony every seven to ten days.
- Wing clip the queen.

When one of the queen's wings is clipped, she is unable to fly. The reasoning behind this is that if the bees swarm they are without a queen and are therefore more likely to return to the hive. Hold her carefully and clip one wing with small, sharp scissors. Be careful of her legs! Ideally, get an experienced beekeeper to show you the procedure and practise on a drone.

Taking a swarm

If swarming does take place, and it will at some time, it often happens in early afternoon on a warm June day. Initially the swarm will settle fairly near to the original colony and will stay there for a time while scout bees are sent to look for new accommodation. Obviously this is the best time to retrieve it before it goes farther afield. About half of the colony will have swarmed, leaving younger bees to take over the original hive.

To take a swarm you will need a straw skep or strong carboard box, a pair of secateurs and a sheet or sack. Protective bee clothing and a smoker or fine water spray will also be required.

If the swarm is clustered on a thin branch, it may be possible to cut the branch and transport the bees a short distance by holding that. If not, it will be necessary to shake the swarm into the collecting box. Ideally there will be two people, one to do the shaking and one to hold the box. Be prepared for the weight! A swarm is heavier than you might think.

If the swarm is in a hedge, cut back enough of the growth to allow the box to be placed *above* the swarm, then use a smoker to direct the bees up into it. As soon as the swarm is in the box, place it upside down on the ground, on a sheet or sack. Place a small wedge on one side to provide an exit. Some bees will go out but if the queen is in the box, they will usually return. If she is not there, they will go out to find her and you will need to start again.

If the swarm is somewhere complicated, such as a church steeple, it is best left to the professionals or those with long experience. The local police station has the telephone numbers of beekeepers they can contact if members of the public report a swarm, but bear in mind that, in law, the swarm becomes the property of the taker.

Hiving a swarm

Move the old hive a couple of metres and turn it to face sideways. Then, put a new brood box complete with floor where the old hive has been. It needs frames of wax foundation so that the workers can get to work immediately. It is also a good idea to place in the centre a couple of drawn combs that have already been worked so that the queen can start to lay straight away. Put the cover board with a feeder containing sugar syrup on top of the feed hole, then place an empty super or eke on this, topped with the roof.

The reason for moving the hives is that bees recognise the position not the hive, so any bees that are flying will return to the new hive, helping to ensure that the new colony continues to be built up. It also reduces the possibility of a second swarm or cast emerging. (See following page).

Hiving the swarm is best done in the evening and the simplest method is to place a piece of wood, the same width as the hive entrance, sloping from the ground to the entrance. Now place a white sheet at the end of this and shake the bees onto the sheet. Some will fly off but most will remain. Bees have a natural tendency to move upwards and they should soon go up the slope and into the hive. Some beekeepers merely shake the swarm straight from the box or skep directly into the brood box. In this case, it helps to put an empty super or eke to act as walls to contain them during the transfer.

Feeding is crucial for the first few days for although the workers will soon begin to forage, there will be insufficient stores, especially if there is a sudden bout of bad weather. At least 4.5 litres (1 gallon) should be made available during the first week. If, after this time, the nectar flow is good and the brood foundation has been drawn into combs, the feeder can be removed. At this stage place a queen excluder over the brood box, with a super above it before replacing the roof.

If the season is a particularly good one and the swarming took place early in the year, there may be a little surplus honey for the beekeeper in the autumn, but in the first season it is best to concentrate on building up a strong colony with adequate stores to go through the winter.

Managing the original colony

Meanwhile, what about the original hive from which the swarm emerged? About half the bees will be left, but there is no queen. However, there will be queen cells with existing sealed brood and some food stores. A virgin queen will emerge approximately 16 days after an egg was laid. She may then proceed to get rid of potential rivals by killing them. If this is the case, she will be fed and cared for by the workers and may take a few practice flights outside the hive. After a few days, when she is mature, she will go out on her mating flight, pursued by the drones, returning after mating has taken place. On her return she will begin to lay eggs in the cells that the workers have prepared for her and the life of the colony continues.

Sometimes, one or more young queens go out of the hive rather than killing the rivals, each taking a swarm with her. These swarms are called casts and are a nuisance to the beekeeper because they are difficult to retrieve and they weaken the colony. Casts can be prevented by destroying all the queen cells except one.

If a cast is retrieved, it is possible to hive it in a brood box with foundation frames, and fed as detailed above. In mild winters it may over-winter successfully, but small numbers of bees find it harder to keep warm. An alternative is to combine it with a stronger colony. (For details see page 63).

An artificial swarm

This is a method of swarm prevention that many beekeepers use and I have always found that it works for me. There are several other methods but it is advisable to do further reading and to talk to experienced beekeepers before trying them. The method I follow is similar to the procedure already discussed on the previous few pages and is used when it is apparent that a colony is about to swarm. Once new queen cells have been sealed, swarming may be regarded as imminent.

Move the hive a couple of metres and place a new brood box with some foundation frames in its position. Make sure that at least one frame has a drawn comb so that the queen can get to work straight away. Leave a gap in the middle of the frames.

Now, go back to the original hive, remove the supers and check the brood comb until you discover the queen. Take out that frame with the queen and any other bees and brood that are on it, and place it in the middle of the new brood box. Destroy any other queen cells that might be on the frame before inserting it. Place a queen excluder and the supers on top of the brood box, followed by the roof.

Workers from the original will now return to the new hive, believing that it is their old one, and will get to work immediately on comb building and bringing in pollen, nectar and water. Any previous conditions that would have led to swarming have now been removed. The original hive colony is dealt with as previously described.

Dealing with an outside swarm

There is always the possibility that you will encounter a swarm of someone else's bees. Having contact with other beekeepers in your area will usually enable you to trace its origin. Although the swarm is legally yours if you take it, nevertheless there is often a tacit agreement that if a beekeeper can identify the swarm as having come from his hive, he should be allowed to come and collect it.

It could be that if you are known to be on the lookout for a swarm, the local society will contact you if there is one available. Reference has already been made to the fact that the police will contact you for swarm removal if you are listed with them.

Once the swarm is hived, some varroa strips should be hung in the hive. (See page 70).

Re-Queening

What is not good for the swarm is not good for the bee.

(Meditations. Marcus Aurelius. AD 160)

Supersedure

Reference has already been made to natural supersedure, where the bees replace the queen with a new one. Artificial supersedure is the replacement of a queen without swarming taking place, and is carried out by the beekeeper. After two years the queen may not be as productive as before and providing a new one reduces the chances of swarming, as well as calming the colony if they are becoming bad tempered.

The replacement queen may be one of your own rearing or she can be bought from a specialist breeder. In the case of the latter, she will usually be supplied in a small cage with a plug of candy at the bottom.

Before the new queen is introduced the old one must be removed otherwise the colony will not accept her. Find her and destroy her. Now take the cage with the new queen and release any attendant bees. Insert the cage with the plug end down between the middle frames of the brood box. After a few days the bees will have accepted her and indeed, the candy plug may well have been eaten so that she will make her own way out into the hive. If not, remove the plug and replace the cage. The queen's pheromones will soon have an effect on the rest of the colony. The cage can be removed once it is apparent that she has left it.

Another, and perhaps easier method, is to introduce the queen into a nuc then, when she is laying, it can be united with the colony.

Marking the queen

It is useful to mark the queen, not only to make it easier to see her during hive inspection, but also to establish her age. There is an internationally agreed system of colour marking as shown:

Year ending in 0 or 5	Blue
1 or 6	White
2 or 7	Yellow
3 or 8	Red
4 or 9	Green

It is quite easy to mark a queen and there are special kits for doing so. A restraining cage is placed over her in the comb, in order to confine her. This has spikes and nylon threads to provide the bars. Hold the frame horizontally, find the queen on it then gently place the cage over her. Take care not to damage her or any nearby bees!

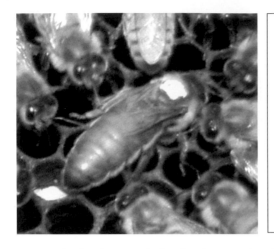

A restraining cage for placing over a queen while she is on the frame so that she can be marked through the bars.

A queen that has been marked on her thorax so that she is easy to identify in future.

When you can see her thorax (the area between head and abdomen) clearly through the bars, place a dab of the appropriate colour on it and remove the cage. It is a good idea to practise the technique on drones first.

There are purpose-made marker pens available from suppliers, or a small dab of queen paint can be applied with a matchstick. Some beekeepers just find the queen, pick her up and hold her while she is being marked, but this is not recommended for beginners, as it is easy to damage her. A steady, patient and gentle touch is required, which come to think of it, sums up quite nicely, beekeeping in general.

Making a nuc

Beginners may have purchased their first bees in a nucleus box that contained about six frames. Once these are transferred to the main hive, the nucleus box is worth keeping for making up your own nuc in future. They are also useful for the temporary housing of a swarm. There are several reasons for making a nuc, including swarm control and increasing one's stock.

Remove four brood frames from an existing hive. These should be from a strong, disease-free colony and have brood and attendant bees but no queen. However, do ensure that there is at least one queen cell. Transfer these four frames to the nuc and replace them with

A travelling box can be used to make your own nuc

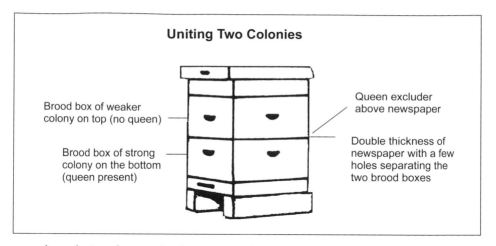

Uniting Two Colonies

Brood box of weaker colony on top (no queen)

Brood box of strong colony on the bottom (queen present)

Queen excluder above newspaper

Double thickness of newspaper with a few holes separating the two brood boxes

new foundation frames in the existing hive. Site the nucleus box away from the apiary. If the nucleus box is a six-frame one, add an empty foundation frame on each side. There will be some food stored on the brood frames, and the workers will bring in forage if the weather is suitable. However, it is still necessary to provide food for the nuc's inhabitants. A small feeder such as one of those detailed in the *Feeding* chapter is appropriate. Equipment suppliers also stock useful entrance feeders that are suitable for a nucleus. Once established and working harmoniously the colony frames can be transferred to a new, full-sized hive. Details of this were given on page 42. The nuc may also be united with another colony.

Uniting two colonies

Uniting two colonies can be done in spring or autumn and there are several reasons for doing it. In spring, for example, a colony may have come through the winter queenless and this can be combined with another colony with a good, young queen. The most common reason for uniting colonies, however, is to strengthen the bees' ability to get through the winter.

When uniting, a weak colony should always be added to a stronger one, never the other way round. Before proceeding, bring the two hives with the colonies together in stages by about a metre each day, giving the bees the opportunity to reorientate themselves. Remember that if one of the colonies was a nuc in a travelling box, it will need to have been transferred to a full-size brood box in a hive a few weeks before.

Moving and uniting is best carried out during the evening. Check before uniting that both colonies have sufficient food stores. If in any doubt, feed before taking action. Also, there may be waxcomb protruding at the top and bottom of the frame. It needs to be cut away and removed beforehand.

Now decide which queen will serve the united colony. With the aid of a smoker, find, remove and kill the other one. Remove the roof and crown board of the stronger colony and lay a double sheet of newspaper across the top of the brood frames. The newspaper should have a few small holes in it, made with the hive tool. Place a queen excluder over the newspaper. Lift the brood box of the queenless colony and place it on top. Now, replace the crown board and roof and leave the bees to sort themselves out.

The scents from the two colonies will combine. The bees will chew through the newspaper and unite. Leave them alone for at least a week before opening up the hive. At this stage, the remnants of the newspaper can be removed and the frames rearranged so that most of the brood is in the centre of the bottom brood box. If necessary, the top brood box can now be removed and replaced with a super.

There are other ways of uniting colonies but this is a relatively simple method that has always worked for me.

Queen rearing

For beginners it is better to purchase a good queen from a reputable supplier. The vitality, health and performance of the colony stem from the qualities of the queen. However, for those who intend to progress as beekeepers, rearing one's own queen is a useful thing to learn. The local bee association can certainly help in this respect, with advice and practical assistance. Rearing new queens provides an opportunity to breed out unwelcome characteristics such as aggression and tendency to swarm frequently but it must be emphasised that although rearing a queen is not too difficult, to breed selectively for required traits is a job for the expert.

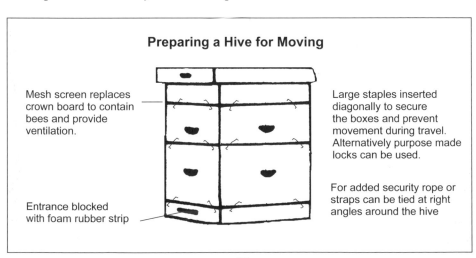

Preparing a Hive for Moving

Mesh screen replaces crown board to contain bees and provide ventilation.

Large staples inserted diagonally to secure the boxes and prevent movement during travel. Alternatively purpose made locks can be used.

For added security rope or straps can be tied at right angles around the hive

Entrance blocked with foam rubber strip

Moving hives

L ike everyone else, beekeepers are prone to back trouble over time so precautions should be taken when moving hives. Always have the hives on a strong stand at least 30cm (1ft) above the ground to reduce the amount of bending, and keep hives in pairs so that when removing supers they can be rested on the next hive. Site hives in such a way that a wheeled vehicle can be brought up close. In the case of a garden apiary, this will probably be a small trolley which is much safer to use than a wheelbarrow as it does not tip. Hives on an outside site need car or van access.

The reasons for moving hives are many and varied. It may be that they are moved short distances during swarm management or for the uniting of two colonies. During the honey harvest, heavy supers will need to be transported. There may be occasions when entire hives are moved from place to place, siting them in orchards, near rape fields or among the heather moors.

Before moving hives examine the chosen site and prepare it so that the hives will be level and entrances unobscured by vegetation. Bear in mind that hives on outside sites have been stolen, so siting them where they are not in full view of a public road is advisable.

Moving a hive in winter is relatively straightforward for there will be little flying about. In summer, it is more complicated, with extra precautions being needed. Moving a hive within an existing apiary should not exceed a metre (3ft) a day so that the bees can orientate themselves each time. Moving a hive away from the apiary means a minimum distance of three miles which is outside the bees' flying range and familiar landmarks. Hives should always be moved at night or early in the morning.

Preparing a hive for moving

First replace the crown board with a mesh screen to provide extra ventilation so that the bees do not over-heat. Needless to say, the mesh is fine enough to prevent any bees getting through it. There is less chance of over-heating when hives are moved in April to the rape fields, but it could be a different story when they are moved in July or August to the heather moors.

Next day, the entrance of the hive should be blocked with a strip of foam rubber which will not slip. The boxes of the hive need to be securely fastened together so that there is no chance of the bees escaping in transit.

Hives sited in the mountains of Spain where the bees can take advantage of the wild flowers of the trees and shrubs, away from the danger of agricultural spraying.

This can be done with thin rope or straps placed at right angles and pulled together firmly. Wood staples can also be inserted diagonally to pin the boxes together and prevent twisting and slippage. There are also special clips available from suppliers, which will lock the boxes together securely.

The hives in transit

Place the hives in the car, van or trailer with the frames aligned with the road in order to minimise vibration. In warm weather have a fine water spray ready so that if the bees show signs of over-heating (a louder buzzing than before) it can be used to spray the mesh screen and help to cool them.

Once set up on a level site, remove the ropes, mesh screen, foam rubber and staples or clips (although some beekeepers leave most of the staples and clips in place, ready for moving the hive back). Place a super and crown board over a queen excluder, unless these were already in position before the hive was moved. Replace the roof.

Retrieving the hives may require a fresh empty super, as the brood box could be insufficient to cope with the number of bees once the supers have been lifted off. The hives will also need to be removed if any chemical spraying is to take place, but farmers in the UK are required to inform beekeepers when they are going to spray, so that there is time to remove them.

Pests and Diseases

For where's the state beneath the firmament
That doth excel the bees for government?
(Guillaume de Salluste du Bartas. 16th Century)

The best way of avoiding problems is to start with sound, clean hives and equipment, have good healthy stock and then practise good management. Carry out thorough inspections in spring and autumn, and regularly replace old brood combs by melting them down and replacing with new foundation. Never buy old combs and always sterilise secondhand hives by scorching them with a blow torch. Even so, there may still be problems in the best run apiary so being prepared is recommended. This includes being a member of a local bee association and attending as many practical demonstrations as possible. In this way you are in the best position to recognise problems. Experienced beekeepers are an invaluable source of help and advice and will also know when to call in the local bee inspector or to advise on other procedures such as taking sample bees for testing.

Adult bee diseases

Nosema

This is caused by a protozoan, *Nosema apis*, in the bee's digestive system. It is not fatal but infected bees are not able to function properly and the colony fails to build up in the way that a healthy one would. Symptoms may include dysentery which stains combs and the front of the hive.

To minimise it, carry out a routine cleaning of empty hives every autumn and sterilise with 80% acetic acid. Beware, for this is corrosive so wear a mask and gloves! The easiest way to apply acetic acid is to use absorbent fume pads (from suppliers). These are placed on the sealed hive floor and saturated with acetic acid before placing the brood box with frames on top. More saturated pads are put on the frames, with another brood box and finally the roof on top. Leave to fumigate for a week.

To treat bees that are affected with Nosema, the antibiotic *Fumidil B* can be added to the sugar syrup that is given in the autumn.

Acarine

This is caused by the mite *Acarapis woodi* in the trachea or breathing tubes. It weakens the bees so that they are unable to fly. Suspect its presence if there are bees with wings held in a 'K' position. There is no official treatment but it is thought that *Apiguard*, a thymol-based gel can be used for it.

Recognising and Dealing with Problems

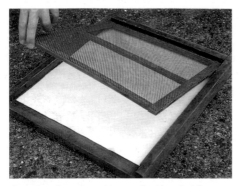

A varroa floor for a hive so that levels of varroa mite can be monitored. The mites fall through the mesh for collection on the floor but cannot return to the hive.

Samples of drone larvae are removed from the brood comb with a drone fork. The dark spots are varroa mites.

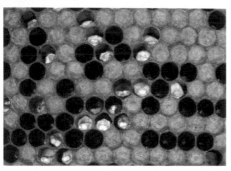

Chalk brood with chalk white 'mummies' of larvae that have been killed by the fungus *Acosphaera apis*.

Sac brood showing 'Chinese slipper' taken from a cell on the affected brood comb.

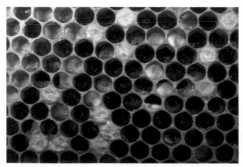

European foul brood affecting unsealed brood.

American foul brood 'ropiness' test.

Photographs by courtesy of DEFRA

Amoeba

This is a protozoan that affects the gut of the bee and tends to affect older bees more than young ones. It often causes dysentery so may be confused with Nosema. Treatment is the same as for that disease.

Paralysis

It is a virus that causes paralysis. Dead bees may be seen in front of the hive, with more dying each day. There is no treatment, but colonies can usually cope as breeding maintains the numbers. Re-queening helps to build up new, healthy stock.

Brood diseases

American foul brood

This is a notifiable disease and if suspected, must be notified to the authorities. It is caused by the spore-producing bacterium *Paenibacillus larvae*. Infected larvae die in the cell and the cappings become sunken and darker in colour. If a matchstick is inserted, the contents are withdrawn as a brownish thread, a procedure known as the 'ropiness' test. The dead larvae gradually dry to dark brown scales. There is no treatment and affected colonies and combs must be destroyed under the supervision of a bee inspector.

European foul brood

This is also a notifiable disease caused by the bacterium *Melissococcus pluton*. It is mainly unsealed brood that is affected, with the larvae dying before the cells are capped. The dead larvae have a melted appearance and the cell contents cannot be withdrawn in the same way as for the AFB ropiness test.
The most effective treatment is to shake the swarm into a new hive, but very weak or reinfested colonies will be destroyed.

Sac brood

This is a very common virus condition for which there is no treatment. It can sometimes be mistaken for AFB so if in doubt, get expert opinion. It affects only sealed brood, where a few larvae turn yellow then dry to a thin dark scale usually called a 'Chinese slipper'. If it persists, it is advisable to re-queen.

Chalk brood

Another common condition, this is caused by the fungus *Ascosphaera apis*. The fungus threads kill the larvae after they have been capped. The larvae appear as chalky white 'mummies'. It is rarely a problem and there is no specific treatment, although weak colonies which are more susceptible can be strengthened by re-queening.

The Difference between Braula and Varroa

Braula coeca,
a harmless and
flightless insect
with six legs,
that is easily visible.

Varroa jacobsoni,
a parasitic, crab-like
mite with eight legs.
It is smaller than Braula
and can be seen as a
reddish-brown dot.

Varroa

This is a parasitic mite that lives on bees and sucks their blood. An infestation which is called varroosis disease is technically a notifiable disease but as practically every colony now has it, the current approach is to adopt an integrated system of management. Varroa first entered Britain in 1992 and has been spreading ever since. A low level of infestation may not affect the colony too much but a heavy infestation can bring about a 'colony collapse'. The emphasis is therefore on checks, prevention, and reduction methods.

• **Recognise them** Varroa mites are sometimes confused with Braula insects, so it is important to be able to distinguish between them. Varroa mites have eight legs. See the illustrations above. A magnifying glass helps.

• **Monitor them** Have an open mesh floor (OMF) or a varroa floor in the hive. About 30% of the mites fall through the OMF mesh but cannot return and the bees cannot get through the mesh.

• **Drone larvae** are particularly targeted by varroa. Regular samples removed with a drone fork for inspection will reveal the mites as brown spots.

• **Treatment** It is important for local beekeepers to treat their colonies at overlapping times in order to minimise mite migration. There are various treatments which are carried out in autumn and spring, and it is important to follow the directions. Varroacide strips are hung between the combs and left for a stated period. They can be used safely during the honey flow but not once the honey supers have been added to the hive. Treatments include *Apistan* or *Bayvarol* strips, and *Apiguard*, a thymol based gel.

In a small book like this there is insufficient room to give details of a fully integrated system of varroa management, and at present we are seeing the arrival of varroa mites that are resistant to some treatments. It is emphasised again that professional help is essential.

Pests

Braula (Bee louse)

Braula coeca is a tiny and harmless insect. It hops like a flea and often clings to a bee's body. It has six legs. Varroa treatment also kills Braula.

Wax moth

These are small, grey-brown moths that tunnel into wax combs to lay their eggs and produce webs of tangled silvery threads. A strong bee colony will promptly expel them before this happens, but a weaker colony is more vulnerable. Bald brood is associated with wax moths. This is where there are abnormal cell cappings with a round hole containing the moth larvae.

Supers stored in winter may contain eggs but as these are vulnerable to the cold, storing the frames in a cold shed should finish them off. Another method is to place the frames in the freezer for 24 hours. There are also proprietary products available from suppliers for their control.

Mice

Field mice may enter a hive in autumn and make a nest, as well as consuming bee stores. The bees find their smell offensive and will not sting them. The best approach is to insert mouse guards on the hive entrance in August so that they are not able to get in.

Woodpeckers

Woodpeckers can cause considerable damage to hives during the winter months, especially if the hives are near woodland. To prevent them doing this, drape plastic sacks over the hives, and weigh them down so they stay in place. Ensure that hive ventilation is not affected.

Wasps

In late summer, wasps can be nuisance by entering the hives and helping themselves to honey. Reduce the size of the hive entrance and if the wasps' nest is found nearby, destroy it.

Ants

These seem to be tolerated by bees if not by the beekeeper. If they cannot be eliminated it is best to move the hives.

Other problems

Pesticide poisoning

Rural beekeepers should keep in touch with their local association who will have been warned by farmers and growers about future sprayings.

Robbing

In late summer, bees from other colonies may try to rob the stores of a weaker one. You can sometimes see fighting between guard bees and the robbers at the entrance. Avoid spilling honey or syrup and clear up any spills that might attract them straight away. Finally, fit a small hive entrance.

Here, the frame's lug is placed in a hole in a supporting bar to keep it steady while the wax cappings are cut from the honey comb with a serrated knife. After cutting the cappings on both sides, the frame is then placed in the extractor for the removal of honey. *(Anna Chambers)*

Honey and other Products

Butter and honey shall he eat.
(Isaiah, The Bible).

Honey is formed from plant nectar or honeydew which is carried back to the hive in the worker's honey stomach. Stored in cells on the honeycomb, it has some of the water evaporated from it by draughts caused by fanning of the wings. Then, it is ripe and ready for capping with wax. This is the colony's winter store, but it is also the beekeeper's harvest so it is removed and replaced with sugar syrup. The honey is ready to be removed when the cells on the frame have been capped.

Removing the honey supers

Before removing capped supers the bees need to be removed from them. As with everything in beekeeping, there are several methods of doing this, but the easiest way is to have one or two Porter bee escape valves inserted in the crown board. These enable bees to pass down from the super into the brood box, but not back again. Leave for 24 hours then check. Most of the bees should have gone down but there will still be some on the frames.

Lift off the roof and super and replace the roof on the hive. Move the super well away from the hive then take out each frame and gently brush off any remaining bees with a bee brush or quill feather before replacing it in the super. The bees will be disorientated but will eventually find their way back to the hive.

Take the super inside to the place where honey extraction is to take place. Depending on the scale, this might be a purpose-built room, a shed or the kitchen. Make sure that all the windows are closed to prevent any stray bees getting in. It is also worth mentioning that a full super is heavy so having a trolley to move it is a real advantage. DIY stores sell them quite cheaply.

Extracting and dealing with the honey

Extracting honey is first and foremost a sticky business. The honey will still be warm when it is taken from the hive and is best extracted from the frames straight away.

There are different types of extractor but the most common are *tangential* and *radial*. The former has the frames placed as shown in the photograph on page 74, while the latter has them resembling the spokes of a wheel. Extractors are available as hand operated models or those that are electrically powered. The sequence for extraction is as follows:

Stages in Honey Extraction

1. When the honey is capped it is ready to be extracted, but first the frames need to be cleared of bees.

3. The uncapped frames are placed in the extractor. This is a tangential type. *(Anna Chambers)*

2. The honey combs are uncapped. *(Anna Chambers)*

4. The extracted honey is removed to a settling tank via the tap at the bottom of the extractor.

5. (Left) After settling it is filtered into storage containers and jars. *(Anna Chambers)*

On a small scale, kitchen utensils can be used for filtering and acting as settling tanks, but a extractor will need to be bought, borrowed or hired. Table-top models are available from suppliers.

The finished product. Honey being offered for sale at a country show, *(Anna Chambers)*

• First place a frame on end, resting over a container to catch the cappings. This could be a deep tray-type container or a food-grade plastic bucket. In the case of the latter, having a wooden bar that fits over the top is an advantage, especially if it has a hole into which the frame lug fits for stability.

• Using a purpose-made capping knife or a serrated kitchen knife, slice off the cappings on both sides and then place the frame in the extractor. Continue until the extractor is full. A warm knife makes the job of uncapping easier so a jug of hot water to dip the knife into is a help.

• As the handle of the extractor is turned (or an electric machine switched on) the honey is flung out against the walls of the outer container, by centrifugal force, so that it collects at the bottom of the container while the wax is contained in the central cage or screen. Rotate the extractor gently at first, gradually increasing the speed. With a tangential extractor it is necessary to remove and reverse the frames so that extraction takes place equally on both sides of the frames. Continue until all the honey is extracted.

• The bottom of the extractor has a tap for removing the honey. As the honey goes out it should go through a coarse filter such as a stainless steel or polythene kitchen sieve, or a straining cloth to catch the bits of wax that get through. A finer filter can then also be used to produce a clearer honey. Double filters with a coarse and a fine screen below it are also available.

Tools for Special Honey

Honey creamer

Comb cutter

Cut section of honey comb

Heather honey press

• Don't forget that the tub of cappings also contains quite a bit of honey that can be filtered and reclaimed. (See Beeswax Page 78).

• If the honey has gone cold it is difficult to filter so a second filtering can be left until later, when it is warmed prior to bottling. If the honey is to be sold, a second filtering is essential but for home consumption you may not be too bothered about still having some wax in it.

• Run the honey into a large container or settling tank and leave to settle for 24 hours. This allows air bubbles to rise to the surface. Keep it covered at this stage. The less dense part of the honey is at the surface and some of the water from it will evaporate. This is the ripening process. Frosting or crystallization often occurs and this can be removed prior to bottling or putting into storage containers.

• If it is to be filtered again at this stage, it will also need warming. On a small scale, the container can be placed in an old domestic boiler or something can be rigged up that will gently warm the honey.

• Glass honey jars are available in 454g (1 lb) and 227g (½ lb) sizes and once they are filled and capped, the honey is ready to be eaten. With a large quantity of honey it is better to store it in large food-grade plastic containers and bottle as necessary.

• Store honey where the temperature does not fluctuate, eg, a pantry.

• Clearing up is the worst part! Rinse everything in cold water to get rid of the debris, then wash in hot water and leave all the equipment to drain.

• The extracted frames can be put back in the super and taken back to the hive in the evening. Remember to remove the bee escape! The bees will then clean the combs for you. If it is fairly late in the season, the supers can then be carefully wrapped and stored away for the winter. The bees are then given their pre-winter feed of sugar syrup.

Oilseed rape honey

In most parts of the country oilseed rape has taken over from white clover as the main source of nectar. The flowering season lasts from May to July. The nectar from rape granulates quickly as it has a high glucose content. Once full, the supers need to be collected promptly. It may even begin to crystallise before capping so this cannot necessarily be taken as a guide to readiness, as with other honey. Many beekeepers in arable areas do not use clearer boards either, relying on brushing away the bees from the supers before taking them in.

The honey needs to be extracted while it is still warm, but if some of the cells are uncapped the frame should be shaken so that any unripe nectar is removed before the rest of the frame is uncapped. Any nectar left will cause fermentation to take place. If it is too late and the honey has set in the frames, it is best to cut out the comb and then melt out the honey.

Apart from heather and acacia honey which do not granulate, most other honey will set after a time, although this can vary from weeks to months. The variation depends on the relative amounts of glucose and fructose sugars in the honey, which in turn, relates to the types of flowers from which the nectar was collected.

Heather honey

Heather honey is popular with customers as it is dark and strongly flavoured. It often sells at a premium price. It also represents an extra harvest for the beekeeper as the heather flowers open in August, just when the flow elsewhere is falling. Heather honey cannot be extracted with an extractor because of its gelatinous consistency. The combs need to be removed and then pressed though a muslin cloth. Heather honey presses are available. An alternative is to use the supers to produce cut comb honey (see below).

Cut comb honey

Cut comb honey is honeycomb where the cappings have not been removed and the honey is still inside. It is popular with connoisseurs. To produce it, the super frames have unwired foundation for the bees to work and produce wax cells for the honey. Once the cells are capped (and here they all need to be capped) the frames are removed and sections of comb are removed with a cutter such as that on page 76. Alternatively, a sharp knife can be used. Small tubs with clear lids are available for selling comb honey.

Creamed honey

Although some people like hard, crystallized honey, not everyone does. A method that can be used to soften it is to cream it. This is done as follows:

Stand the container of granulated honey in hot water (49°C maximum) until it about a third of it has liquified. Now immerse the creamer tool (shown on page 76) and move it up and down, keeping it below the surface to avoid drawing in air. If a creamer tool is not available, stirring well with a tablespoon is an alternative. When the honey takes on an opaque appearance and is quite thick, it can then be bottled and will keep its soft consistency.

There are those who claim that the flavour of creamed honey is not as good as set honey, but this is an aspect on which people must agree to differ. Its shelf life is increased and this method is used to produce shop honey.

Selling honey

Anyone selling honey will need to comply with the relevant legislation. (See page 91). There is also an excellent advisory leaflet on the subject that is available from BBKA. (See page 94 for contact details). As referred to earlier honey destined for sale must be filtered so that it is clear of impurities. It should be in glass jars of appropriate size, with a label that indicates country of origin, the net weight of honey and details of the producer.

Beeswax

Wax is a valuable commodity and as much as possible should be reclaimed. There are several ways of doing this, but first the cappings that were removed before extraction will still contain honey. Put them in a muslin bag and hang this up to allow the honey to drain out. Now put the cappings into a food-grade plastic bucket and cover them with water. Stir it well then strain the liquid into a container where it can be used to make mead later. Squeeze the moisture from the wax cappings and leave them to dry.

Wax can also be reclaimed by putting pieces in a muslin bag and placing this in a large pan of water. When heated gently the wax melts at 62°C and rises to the surface, leaving any impurities in the bag. Even so, there is often discoloration of the wax and an alternative is to use a solar extractor.

Solar extractors are available from bee suppliers or you can make your own. It consists of an insulated box with a double-glazed glass or clear, rigid plastic top at an angle of 45°. Inside is a sloping metal container with another container into which the wax drains. When placed facing south on a sunny day, it is amazing how quickly the wax melts. The resulting wax can be used again for making foundation or for turning into candles or furniture polish. Foundation and moulds are available from suppliers.

Beeswax polish   Put 450g (1lb) of beeswax in a metal container and melt it over a container of hot water. When melted, extinguish all flames and stir in 1 litre (2pt) of pure turpentine, not the substitute. Mix well, pour into containers with lids and label. If sold, it must also have a hazard warning.

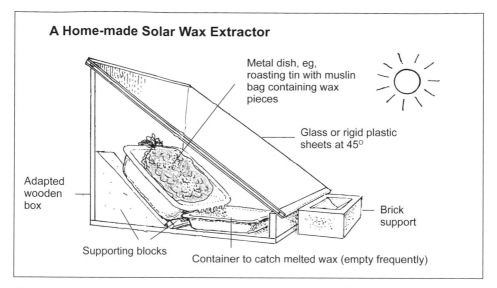

A Home-made Solar Wax Extractor

Metal dish, eg, roasting tin with muslin bag containing wax pieces

Glass or rigid plastic sheets at 45°

Adapted wooden box

Brick support

Supporting blocks

Container to catch melted wax (empty frequently)

Candles Moulds and wicks for making beeswax candles are available from bee or craft suppliers. The process is quite straightforward and after inserting the wick and pouring in the melted wax, the only thing to watch out for is to make sure that the wick is in the centre. When cold, remove the candle from the mould and it is ready for use.

Mead

Mead is essentially a fermented mixture of honey in water. It is advisable to use soft rain water but this is not absolutely essential. The proportion of honey to water can be varied: less honey produces a drier mead while more gives a sweeter, fuller-bodied wine. If you want to use the water left after washing the cappings, strain it well through muslin and test the specific gravity with a hydrometer (available from winemaking suppliers). The specific gravity should be between 1.06 and 1.12. More honey and water can be added as required. The basic recipe is as follows:

Dissolve 1.3kg (3lb) honey in 4.5 litres (1 gallon) water. Add a teaspoonful of yeast nutrient, two tablespoons of cold tea and the juice of one lemon. Sterilize the liquid by bringing it to the boil and allow to simmer for a few minutes. Skim off the scum, cover and leave it to cool.

When it is around 21°C, add a sachet of wine yeast and pour into demijohns fitted with airlocks. Place in the airing cupboard while it ferments. As this declines, top up with boiled, cooled water to maintain the small space below the airlock. When fermentation has ceased, siphon off the mead and discard the deposit at the bottom. Wash and dry the demi-johns, add a Campden tablet and refill with the mead. Store in a cool, dark place. If more deposit forms, repeat the last process. Bottle after one year. Cheers!

An observation hive at a country show.
Here the queen is being pointed out to the little girl and her mother.
(Anna Chambers)

Bees at the Show

The farmers crowd to the fair today in obedience to the same ancient law,
As naturally as bees swarm and follow their queen.
(Henry David Thoreau. 19th Century)

Country shows, whether they be at small village fetes, large agricultural events, county shows or specialist gatherings, are one of the delights of rural life (and increasingly of urban life). All manner of livestock and produce are on show, amongst them bees, honey, beeswax candles and myriad other bee items of abiding interest.

The observation hive
An observation hive is often to be seen at events where there is a beekeeping area. Set up in such a way that there is no public hazard, it has a large, glass observation window. The queen, drones and worker bees can be seen on the combs and it is an excellent way of introducing bees to children and those who may never have seen honeybees close up before.

An experienced beekeeper will be in attendance, ready to point out different aspects of the hive and its occupants. He or she will also be ready to answer questions and pass on information about the craft and its relevant organisations. It is an excellent introduction and is recommended to anyone who wishes to learn a bit more about the world of bees.

Showing honey
Honey and other bee products compete at shows organised by the bee associations. If you intend to have a go at showing your produce get a show schedule from the secretary in good time. This will provide all the information you require. It is worth adding that the best way of learning and seeing the show from the inside, is to offer to help in the bee tent.

There are three levels of shows and it is recommended that you start with the novice class at a local show. From here proceed to a county show, followed by a national show. The most prestigious is the National Honey Show held in London every autumn. Judges who are examining honey exhibits are looking for the following:

- *Correct for class:* Dark honey should not be in the light honey class, etc.
- *Cleanliness:* Well strained to remove impurities. No scum on surface.
- *Clarity:* No haziness or air bubbles.
- *Density:* It has a good 'flow'.
- *Colour:* Good colour for its type.
- *Aroma:* It has a good 'bouquet'.
- *Flavour:* This says it all. Good luck!

Plants such as the buddleia not only provide nectar for bees but also attract butterflies such as this Peacock into the garden.

Teasel is easy to grow from seed and is a popular source of forage for bees in the summer. In the autumn its seeds provide food for garden birds and in winter its seed heads can be dried for decoration.

Fruit trees such as this apple rely on bee and other insect pollination in order to bear fruit.

The Bee Garden

The moan of doves in immemorial elms
And murmuring of innumerable bees.
(The Princess. Tennyson. 1847)

Few people will have a garden that is entirely given over to bees, but most gardens will have room for flowering plants that provide food for honeybees. In fact, with a little planning it is possible to have plants to provide some forage over the whole season. The most important sources of pollen and nectar are fruit trees, willow, lime and heather. In recent years oilseed rape has also become the most prominent fodder crop in arable areas.

The following is a list of trees, shrubs, herbaceous plants and bulbs that are particularly popular with bees. (They will also help to attract butterflies into your garden). The common name and the Latin name are given for each species, but there are often different varieties, too. For example, *Tilia* is the name for the Lime tree, but there are different varieties such as *Tilia maximoweziana* or Japanese Lime, and *Tilia platyphyllos* which is the Broad Leaved Lime. The plants selected will obviously depend on the size of garden that is available and on the type of soil. No trees should be planted too near a dwelling or road in case the foundations are affected by the roots. Some plants are normal garden plants, while others are wild species. These should obviously not be taken from the wild, but are available from specialist suppliers and seed companies. Remember that some plants are invasive and may not be a good choice for a garden. Double-flowered cultivars are not as productive as the old single flowered varieties.

Trees

Fruit trees All fruit trees, shrubs and soft fruit - apple, pear, plum, apricot, currants, gooseberries, strawberries, raspberries, etc, are visited by honeybees. In fact, some commercial orchards have agreements with local beekeepers to site their hives there during the flowering period so that there is a good level of pollination. Crab apples that flower from April to May are also popular.

Alder *Alnus sp.* Up to 19m. Flowering February to April. Pollen only.

Handkerchief Tree *Davidia involucra.* Height 12-19m. May.

Hawthorn *Crataegus sp.* Grows to around 6m. May onwards.

Hazel *Corylus sp.* January to March. Pollen only.

Horse Chestnut *Aesculus sp.* Height 20-30m. May onwards.

Lime *Tilia sp.* The lime is one of the most important sources of bee fodder, but is for large sites only. Height up to 39m. June to August.

Maple *Acer sp.* Height from 4m-25m. Can be invasive. Late April onwards.

Poplar *Populus sp.* Height 30m. February to March. Pollen only.

Pussy Willow *Salix caprea.* 10m. March. Useful when not much else is available.
Rowan (Mountain Ash) *Sorbus aucuparia.* 15m. May to June.
Sweet Gum *Liquidambar styraciflua.* Up to 30m. May onwards.
Tulip Tree *Liriodendron tulipifera.* Up to 55m. June to July.

Shrubs

Bramble Hedgerow rambler. July onwards.
Broom *Cytisus sp.* 1-3m. April to July.
Buddleia (Butterfly Bush). *Buddleia sp.* Around 3m. July to October.
Buckthorn *Rhamnus catharticus.* Hedging but can be invasive. May to July.
Heather *Calluna* and *Erica sp.* Both Ling, *Calluna vulgaris* and Bell Heather, *Erica cinerea,* are important bee crops. They require acid soils. Flowering between July and September, they can grow to a height of 20-75cm.
Holly *Ilex aquifolium.* Up to 4 metres. May to June.
Honeysuckle *Lonicera periclymenum* Climber. June to September.
Hyssop *Hyssopus officinalis.* Herb. 60cm. July to August.
Lavender *Lavendula sp.* Herb. 15-60cm, depending on the variety. July onwards.
Mahonia *Mahonia aquifolium.* 2 metres. March to April.
Sage *Salvia officinalis.* Herb. 70cm. July to August.
Thyme *Thymus serpyllum.* Herb. 60cm. June to July.
Wild Clematis *Clematis vitalba.* Climber. June to August.

Herbaceous plants

Alkanet *Anchusa azurea.* Up to 90cm. June to August.
Alyssum *Lobularia maritima.* Hardy annual. 15cm. June to September.
Angelica *Angelica officinalis.* Biennial. Up to 3 metres. July to August.
Arabis *Arabis caucasia.* Rockery perennial. 30cm. April to May.
Aubretia *Aubretia deltoides.* Rockery perennial. 15cm. April onwards.
Balm (Bee Balm) *Melissa officinalis.* Perennial. 60cm. Can be invasive. Aug.
Black Medick *Mediago lupinus.* Wild annual. 15-30cm. May to October.
Borage *Borago officinalis.* Hardy annual herb. 75cm. May to September.
Candytuft *Iberis sempervivens.* Annual. 20cm. May to August.
Catmint *Nepeta sp.* Rockery perennial. 15cm. May to September.
Chicory *Chicorium intybus.* Vegetable. 45cm. July to August.
Chives *Allium schoenoprasum.* Perennial herb. 30cm.
Coltsfoot *Tussilago farfara.* Wild perennial. 10cm. February to March.
Common Mallow *Malva sylvestris.* Wild biennial. 150cm. June to October.
Cornflower *Centaurea montana.* Wild annual. 60cm. May to October.
Corn Marigold *Chrysanthemum segatum.* Wild annual. 60 cm. June onwards.
Corn Poppy *Papaver rhoeas.* Annual. 30cm. July to August.
Dandelion *Taraxacum officinale.* Invasive weed. 20cm. Major source early April.
Elecampane *Helianthum inula.* Perennial. 2 metres. July to August.
Evening Primrose *Oenothera biennis.* Wild biennial. 120cm. June to Sept.
Field beans *Favus sp.* Important field crop if not sprayed. June to July.
Flax *Linaria purpurea.* 15cm. June to August.

Forget-me-not *Myosotis sp.* Hardy annual. 25cm. April and October.
Golden Rod. *Solidago sp.* Perennial. 120cm. July to September.
Globe Thistle *Echinops ritro.* Perennial. 120cm. July to October.
Hardy Fuchsia *Fuchsia sp.* Perennial. Up to 1.5 metres. March to October.
Himalayan Balsam *Impatiens glandulifera.* Invasive annual. 3m. July to Oct.
Honesty *Lunaria biennis.* Biennial. Up to 90cm. April to June.
Hollyhock *Althea rosea.* Perennial. 3m. July to August.
Lobelia *Lobelia erinus.* 10cm. Half hardy annual. June to September.
Lupin *Lupinus polyphyllus.* Perennial. 90cm. June to July.
Marjoram *Origanum vulgare* Perennial herb. 60cm. July to September.
Masterwort *Astrantia major.* Perennial. 60cm. June to July.
Meadow Cranesbill *Geranium praetense.* Wild perennial. 1m. June-Sept.
Meadowsweet *Filipendula ulmaria* Wild perennial. 120cm. June to Sept.
Michaelmas Daisy *Aster novi-belgii.* Perennial. Up to 90cm. September.
Mignonette *Reseda odorata.* Hardy annual. 30cm. July to September.
Nasturtium *Tropaeolium sp.* Annual. 15cm. June to October. Also climbers.
Poached Egg Plant *Limnanthes douglasii* Hardy annual. 15cm. April onwards.
Primrose *Primula sp.* Perennial. 10-35cm. April to May and in the autumn.
Red Clover *Trifolium pratense.* Meadow flower. 30cm. May to October.
Self-Heal *Prunella vulgaris.* Wild annual. 20cm. June to October.
Star of the Veldt *Dimorphotheca aurantiaca.* Hardy annual. 30cm. June-Aug.
Stonecrop (Ice Plant) *Sedum spectabile.* Perennial. 30-60cm. Summer-autumn.
Sunflower *Helianthus annus.* Annual. Up to 3 metres. July to September.
Sweet Cicely *Myrrhis odorata.* 60cm. May to June.
Sweet Violet *Viola odorata.* Perennial. 5-20cm. June to September.
Teasel *Dipsacum fullonum.* Biennial. Up to 2 metres. July to August
Tree Lupin *Lupinus arboreus.* Perennial. 120cm. June to September.
Valerian *Centranthus rubra.* Invasive perennial. 30-80cm. May to September.
Viper's Bugloss *Echium vulgare.* Wild annual. 30cm. June to August.
Wallflower *Cheiranthus cheiri* Biennial. 20-60cm. March to June.
Welsh Poppy *Meconopsis cambrica.* Perennial. 45cm. June to July.
White Clover *Trifolium repens.* Important meadow flower. 30cm. May to October.
Willow Herb *Epilobium sp.* Wild annual. Up to 1 metre. June to August.
Woad *Isatis tinctoria.* Biennial herb. 90cm. June.

Bulbs and tubers
Aconite *Eranthis hyemali.* Perennial. 10cm. January to March.
Bluebell *Endymion sp.* Perennial. 15cm. April to May.
Crocus *Crocus sp.* Perennial. 10cm. March to April. Also autumn varieties.
Grape Hyacinth *Muscari sp.* Perennial. 15cm. March to April.
Snowdrop *Galanthus nivalis.* Perennial. 15cm. January to March.
Snowflake *Leucojum vernum.* Perennial. 30cm. February to May.
Squill *Scilla verna.* Perennial. 10cm. February to May.
Solomon's Seal *Polygonatum multiflorum.* Perennial. 60cm. May to June.

The Beekeeper's Year

Crowds of bees are giddy with clover.
(Divided. Jean Ingelow. 19th Century)

Many of the procedures that are required in beekeeping have already been covered earlier in the book, but in this chapter, it is worth drawing them all together in a general survey of the beekeeper's year. Please note that this provides general guidance only, for much depends on the weather at any particular time. In a very mild spring, for example, the bees may emerge and start working much earlier than they did the previous year. Unfortunately, the opposite is also true!

Spring
Spring unlocks the flowers to paint the laughing soil.
(Seventh Sunday after Trinity. Reginald Heber. 19th Century)

March
Although some bees will have emerged earlier in the year on cleansing flights to get rid of droppings, more of them are now emerging to investigate the blackthorn and wild cherry in the hedgerows. Alders, poplars and pussy willows are also in flower, with gardens providing bulbs such as aconites, crocus and grape hyacinths. Even so, there may be a shortage of food as more brood is laid and reared. This is a dangerous time when it is possible to lose colonies from starvation. Check the weight of the hives and, if necessary, feed with sugar syrup. In a cold, late spring it may also be necessary to feed pollen or a pollen substitute.

If, despite feeding, the bees appear sluggish, suspect disease. Take a sample of bees and have them tested. The first drones will now be apparent.

On a mild day take a brief look inside the hive and begin a monitoring programme for varroa in association with the local bee association.

April
Now is an ideal time for starting in beekeeping if you have not kept them before. The best way is to buy a nuc from a supplier. This will have four or six frames, a young queen, brood, food and some attendant workers. The frames can then be transferred from the travelling box into a hive.

It is the start of the main spring flow with fruit trees, crab apples and maples, as well as honesty, primroses, poached egg flowers and forget-me-nots. The garden rockery provides arabis and aubretia, while lawns, roadsides and banks are resplendent with dandelions.

Choose a mild day for the first major inspection. (In mild areas this may have been at the end of last month). Check that the queen is present and that there is no sign of disease. Mark the queen for future identification if she is unmarked. Check the availabity of food and feed if necessary. Provide new brood foundation combs as needed.

At the middle or end of the month put on a queen excluder and a super with frames of foundation.

May

May brings the soft fruit, hawthorn and horse chestnut, although the most plentiful supply of nectar and pollen is now oilseed rape. If it is grown nearby arrange to move some hives there, if required.

The queen will be at her peak of egg laying and there should be plenty of brood in the brood box. If there is insufficient room, provide another brood box or a super on top of the first one. Replace the queen excluder and honey super above these.

Keep a lookout for swarming preparations such as the building of queen cells and take appropriate action. Artificially swarm if this is needed.

Prepare a spare hive in case swarming does occur and have the necessary equipment ready for taking it.

If planning to expand the stock of bees, make up your own nucs.

Set up a solar wax extractor to reclaim bits of wax.

Clean and scrape hive floor boards.

Summer

My banks they are furnished with bees whose murmur invites one to sleep.
(A Pastoral. William Shenstone. 18th Century)

June

If the bees have been working an oilseed rape crop, put clearer boards on the hives, or insert bee escapes in the crown board. The honey must be taken quickly before it granulates on the frames. Keep the supers warm and extract the honey immediately.

Field beans, white clover and lime trees are now available for foraging. Check that agricultural fields are not about to be sprayed.

Continue to carry out weekly checks for indications of swarming.

July

The queen's rate of laying is now beginning to decline. Continue checking for swarming activity each week. If the nectar flow is particularly good add extra supers as needed. Hedgerow brambles and meadow sweet are now in flower, as well as many garden plants such as lupins and mignonette.

August

Unless you are in an arable oilseed rape area, from now to September is the time for the main honey harvest. Have all the necessary extracting and storage equipment ready in good time.

Arrange to move hives to the heather moors if required.

It should no longer be necessary to check for signs of swarming.

If you have a newly reared queen she can replace the old queen. If the bees have been showing signs of being bad tempered, this will have a calming effect.

Make the size of the entrance smaller to make robbing by wasps less likely.

Autumn

I saw old Autumn in the misty morn stand shadowless like silence,
Listening to the silence. (Autumn. Thomas Hood. 19th Century)

September

Continue to take honey, as available, if not harvested in the previous month. The queen will usually stop laying at this time, and the drones will also disappear. Any left in the hive will be driven out by the workers.

Treat for varroa as soon as the last honey has been harvested.

Feed the bees for the winter so that they have time to take the sugar syrup down and cap it for their winter stores.

October

The bees are showing less activity outside the hive now. Feed for the winter but no later than the middle of the month. Put mouse guards at the hive entrances and check that the ventilation is not restricted.

Winter

This is the time of hanging on for the bees - the bees
So slow I hardly know them. (Wintering. Sylvia Plath. 1962)

November

As soon as it gets cold the bees begin to cluster in a ball for warmth. They are best left alone without opening and disturbing the hive.

Now is a good time to provide some weights on the roof of the hive, or to tie it down, in case of high winds. Clear away any debris or long growth under the hive supports. Repair fences, as needed, and begin to prune trees and shrubs that require it. Read some bee books and relax.

December

The bees will still be clustered in a ball to keep warm and should not be disturbed. Get catalogues from the bee equipment suppliers and plan next season's work. Check equipment and see if anything needs to be repaired or replaced. Make some beeswax candles for Christmas and also enjoy some of last year's mead.

January

The bees are still clustered and should not be disturbed. Carry on with pruning and clearing on frost-free days, and make regular checks for wind damage such as fallen branches, etc. If there has been

Taking a sample of bees for testing. (*Anna Chambers*)

snow, shield the entrance to the hive so that bright light is not reflected inside. If it does, it may attract bees outside and they will be killed by the cold. Try making some wax foundation with a purpose-made mould from suppliers. Now might be a good time for a holiday in the sun!

February

The queen begins to lay so it is important that there is enough food for the colony. Fondant is better at this stage because sugar syrup excites them and may make them leave the hive when outside conditions are unsuitable.

On mild days some bees do emerge on cleansing flights to get rid of droppings. They may also be attracted to early pussy willows and coltsfoot on banks and hedges, and by snowdrops in the garden. An important task at this time is to make a source of water available to them so that they do not have to fly a long distance to find it.

Appendix I: Bee Stings

A bee commits hara-kiri every time it bayonets you with its little sting.
(Sellar and Yeatman, Garden Rubbish.1936)

There are many misconceptions about bees and bee stings. The drone has no sting. The queen has one but uses it only to sting and kill off queen larvae that pose a threat to her existence within the hive. She retains the ability to use her sting several times. This is unlike the worker which dies after it has stung. This is because there is a barb on the stinger which cannot be withdrawn without causing irreparable damage to the abdomen.

As far as the worker bee is concerned, her sting is a defensive mechanism that is used to protect the hive and its inhabitants. It is not in her interest to waste the venom and commit suicide without good reason. People who are nervous of bees probably do not think about the fact that there are many thousands of bees which visit their gardens every year, yet they are not stung. It is when they are in the proximity of the hive that the workers are most likely to sting, hence the need for protective clothing when opening the hive to examine the frames. The point was made earlier in the book that there are some people who are allergic to bee stings and it may not be a good idea to have hives in their gardens, but it is obviously a personal decision that should be based on medical advice.

It pays to avoid trouble. Bees do not like thundery, muggy weather so it is best not to open up the hive in these conditions. Don't wear dark or furry clothes because, as far as the bees are concerned, you could be a bear! They don't like the smell of sweat either. It is most important to open up the hive carefully so that bees are not crushed. The smell of this upsets the rest of the bees and can make them aggressive.

If in spite of all precautions you do get stung, avoid trying to squeeze out the sting because it will just discharge more toxin into the body. Scrape it off with the hive tool or a fingernail and apply a little disinfectant. Lemon juice or honey are also said to be beneficial but in my experience, a cold compress and a couple of anti-histamine tablets are the most effective if the sting is really swollen and irritating.

Finally, it is worth mentioning that other topic which always seems to spring to mind when stings are mentioned - killer bees! African honey bees are aggressive, attacking quickly and pursuing victims over distance. They are also excellent honey producers which led to their being imported into Brazil in the 1950s. They escaped and by the 1980s had spread south to Argentina and north to Mexico, arriving in the southern states of the USA in the 1990s. Fortunately for us, even in the unlikely event of their introduction, they are unlikely to survive in northern Europe because they do not normally fly in temperatures below 14°C.

Appendix II: Legislation

Health

Both European Union and National legislation apply to the keeping of bees. Full information is available from DEFRA.

Bees Act 1980 (UK) and The Bees Act 1980 (EU)

These empower Agriculture Ministers to make Orders for controlling pests and diseases affecting bees, and also give powers of entry and inspection to authorised people. So far, the following Orders have been made:

The Bee Diseases Control Order 1982
- This designates the following diseases as notifiable, ie, they must be reported to the authorities: American Foul Brood, European Foul Brood and Varroosis.
- If required, the beekeeper must provide information about the number and whereabouts of the colonies and allow access to bee inspectors.
- Inspectors have the right to examine hives and take samples, mark hives for identification purposes, destroy colonies infected with AFB or EFB, and treat colonies infected with EFB with antibiotic.
- The beekeeper must not move any hives, bees or equipment from a place with suspected brood disease until samples have been sent to the National Bee Unit and declared clear, or inspected by a bee inspector on site who has declared that the brood is free of disease.
- The beekeeper must destroy or treat with antibiotic any infected colonies, under the supervision of an inspector, within 10 days of diagnosis.
- No combs, honey or bee products must be removed from colonies treated with antibiotic within 8 weeks of treatment.
- No substance may be used that disguises the presence of AFB or EFB.

The Importation of Bees Order 1997
This prohibits the import of bees and bee pests except under licence.

Sale of Products

The following legislation applies. Full details are available from the local Environmental Health Office and the local Trading Standards Office. The BBKA also have good guidelines.

- Food and Drugs Act 1984
- Food Hygiene Regulations 1970
- Labelling of Food Regulations 1984
- Honey Regulations 1976
- Materials in Contact with Food Regulations 1978
- Weights and Measures Acts 1963-1979
- Trades Description Acts 1968 and 1972
- Consumer Safety Act 1978
- Labelling Requirements. 2004

Glossary

Acarine	Disease where *Acarapis woodi* mites affect breathing tubes of adults.
AFB	American Foul Brood, a bacterial disease affecting larvae.
Antennae	Sensory feelers on top of the head.
Apiary	Group of two or more bee hives.
Artificial swarming	Technique to prevent swarming, increase stocks or raise new queens.
Bait hive	Brood box with old frames to entice a swarm to enter.
Bee dance	Movements used to communicate information about food sources.
Bee escape	One-way exit that prevents re-entry to the supers, eg. Porter escape.
Bee space	8mm space between frames and hive walls.
Bee wall	Traditional construction for protection of straw skeps.
Bottling tank	Tank from which honey is put into jars.
Brace comb	Comb built between main combs to join them or fill in gaps.
Braula	Bee louse, *Braula coeca*, found on adult bees.
Brood	Eggs, larvae and pupae, either unsealed or sealed (capped with wax).
Brood box	Chamber where brood frames are housed for the queen to lay eggs.
Cappings	Wax covering over brood cells and honey cells.
Caramelisation	Caused by over-heating of honey, producing a burnt sugar taste.
Cast	Swarm with new queen after main swarm.
Cell	Hexagonal space in frame.
Chalk brood	Disease of brood caused by the fungus, *Acosphera apis*.
Chilled brood	Brood that has died from cold.
Cleansing flight	Leaving hive to deposit droppings after periods of confinement.
Clearer board	Crown board with bee escapes, used to clear bees from honey supers.
Cluster	Bees clinging together, either in a swarm or in a hive to keep warm.
Colony	Existing community of bees.
Commercial	Type of hive with large frames.
Crown board	Inner cover under roof of hive.
Dadant	Type of hive with very large frames.
Drawn comb	Comb built out to form full size cells from wax foundation.
Drone	Male bee
Dummy board	Board, used to reduce size of brood box.
Dysentery	Droppings inside the hive, usually the result of disease Nosema.
EFB	European Fowl Brood, a disease affecting larvae.
Eke	Shallow hive box with walls but no roof or floor.
Emergence	When young bees emerge from the brood cells.
Entrance block	Strip of wood to control size of hive entrance.
Enzyme	A natural chemical, eg, invertase that speeds up a chemical reaction.
Extractor	Machine for separating honey from combs using centrifugal force.
Feed hole	Space on the crown board through which bees can feed.
Fondant	Soft sugar candy often used as winter food; also called candy.
Forage	Plants that act as food sources for the bees
Foundation	Wax or plastic sheet on which hexagonal pattern of cells is imprinted.
Frame	Wooden (or plastic) frame for holding wax comb.
Frosting	White appearance of honey which is harmless but spoils presentation.
Granulated honey	Where the honey has set naturally.
Hefting	Assessing amount of food in the hive by checking its weight.
Hive	Artificial environment provided for a bee colony.
Hive tool	Metal tool for levering hive and separating frames.
Hoffmann	Type of self-spacing frame.
Honey	Sweet substance produced from concentrated nectar.
Honeyflow	The period when nectar is available for collecting from flowers.

Honeydew	Sweet secretion of aphids.
Honey sac	Also called honey stomach, where nectar and water are carried.
Hypopharyngeal	Glands where food for feeding young bees is produced by workers.
Invertase	Enzyme that converts sucrose sugar to glucose and fructose.
Langstroth	Type of hive, most popular worldwide.
Larva(e)	Grub stage of bee development.
Mead	Wine produced from honey.
Mouse guard	Device for excluding mice at hive entrance.
Nasonov glands	Glands on the abdomen whose secretion acts as a directional guide.
National	Type of hive, most popular in Britain.
Nectar	Sweet liquid in flowers to attract insects, from which honey is made.
Nosema	Disease caused by protozoan, *Nosema apis*, affecting gut of adult bees.
Notifiable diseases	Diseases that must, by law, be reported to the authorities.
Nuc	Short for nucleus, a small colony of bees, around four frames in size.
Nuptial flight	Flight of virgin queen when she is mated by accompanying drones.
OMF	Open Mesh Floor.
OSR	Oil seed rape, biggest single source of nectar in UK.
Parthogenesis	Virgin birth resulting from unfertilised eggs, as in the case of drones.
Pheromone	Hormone to communicate information, also called queen substance.
Pollen	Male cells of flowers, and the main protein food of bees.
Pollen basket	Also called pollen sac, where bees transport pollen to the hive.
Propolis	Sticky substance from tree buds, also called bee glue.
Pupa	Resting or chrysalis stage when bee grubs are sealed in the cell.
Queen	Sexually developed female bee that lays eggs in the colony.
Queen excluder	Screen which allows workers to pass but excludes queen and drones.
Queen cell	Cell in which a queen bee is reared.
Queen cup	First stage of a queen cell, resembling an acorn cup.
Ripening	Evaporating and concentrating nectar or sugar solution for storage.
Robbing	Stealing of food stores by other bees or wasps.
Royal jelly	Food secreted by worker bees for feeding the queen.
Sac brood	Virus disease affecting brood cells.
Settling tank	Where honey settles to get rid of air bubbles after extraction.
Skep	Old form of hive made of straw, etc, now often used to take swarms.
Smith	Type of small hive.
Smoker	Device for producing cool smoke to pacify bees when hive is opened.
Spiracles	Breathing holes on an insect's body.
Sucrose	Granulated cane or beet sugar.
Super	Part of hive above brood box, where honey is stored.
Supersedure	Process of replacing the queen with a younger one without swarming.
Swarm	Group of bees that have left hive with a queen to form a new colony.
Thorax	Central area of an insect's body between the head and abdomen.
Uniting	Combining two or more bee colonies.
Varroa	Parasitic mite, *Varroa jacobsoni*, that causes varroosis disease.
Veil	Protective head covering used when handling bees.
Venom	Toxin produced by bee during the process of stinging.
Virgin queen	Young and unmated queen.
Viscosity	The flowing capability of honey.
Warming cabinet	Insulated and heated container for liquefying honey.
Wax	Substance produced by worker bees for building combs.
Wax glands	Glands on underside of bee's abdomen for secreting wax.
Wax moth	Insect whose larvae destroy wax comb.
WBC	Type of hive with outer cover lifts to protect inner bee boxes.
Worker	Under-developed female bee that carries out the work of the colony.

Reference section

Books

A Handbook of Beekeeping. H R C Riches. Northern Bee Books.
Background to Beekeeping. A C Waine. Bee Books New and Old.
Bees at the Bottom of the Garden. Alan Campion. Trafford Press.
Guide to Bees and Honey. Ted Hooper. Marston House.
Honey Bees: A Guide to Management. Ron Brown. Crowood Press.
Practical Beekeeping. Clive de Bruyn. Crowood Press.

Bee Book Suppliers

Bee Books New and Old. Tel: 01432 840529. www.honeyshop.co.uk
Northern Bee Books. Tel: 01422 882751. www.beedata.com

Magazines

Bee Craft. BBKA monthly. £18/year. Tel: 01372 451891. www.bee-craft.com
Beekeepers Quarterly. Northern Bee Books. £25. www.beedata.com
Bee Culture. E.H.Thorne monthly £25/year www.thorne.co.uk
(There are also magazines by beekeeping groups - see Organisations).

Organisations There are local organisations throughout the UK, offering advice and support. They publish newsletters and have regular meetings. Normally affiliated with the BBKA or the SBKA (sec below) and offer an insurance service. Find your local association through the BBKA, SKBA or Thorne's websites below.

BBKA British Beekeepers' Association. Tel: 02476 696679. www.bbka.org.uk
SBKA Scottish Bee Keepers' Association. www.scottishbeekeepers.org.uk
WBKA Welsh Beekeepers' Association. Tel: 01974 298336
UBKA Ulster Beekeepers' Association. www.ubka.org
FIBKA Federation of Irish Beekeepers. www.irishbeekeeping.ie
CABK The Central Association of Bee Keepers. www.cabk.org.uk/index.htm
NBU National Bee Unit. Part of DEFRA includes testing service for bee diseases Tel: 01904 462510. www.csl.gov.uk/science/organ/environ/bee
BIBBA Bee Improvement Bee Breeder Association. www.angus.co.uk/bibba/
IBRA International Bee Research Association. www.ibra.org.uk/
BFA Bee Farmers' Association. www.beefarmers.co.uk
National Honey Show. www.honeyshow.co.uk

Suppliers

E.H. Thorne (Beehives) Ltd. Tel: 01673 858555. www.thorne.co.uk
National Bee Supplies. Tel: 01837 54084. www.beekeeping.co.uk
The Honeyshop. Tel: 01458 253098. www.honeyshop.co.uk
B.J. Sherriff. Tel: 01872 863304. www.bjsherriff.co.uk
Michael Jay Beekeeping Supplies. Tel: 01761 452344. www.beebitz.com
BBWear. Tel: 01872 273693. www.bbwear.co.uk
Maisemore Apiaries. Tel: 01452 700289. www.bees-online.co.uk
Stamfordham Ltd. Tel: 01661 886219. www.stamfordham.biz/
Park Beekeeping Supplies. Tel: 020 8694 9960. www.parkbeekeeping.com
John Gower Bee Supplies. 01452 740719
Paynes Southdown Bee Farms Ltd. Tel: 01273 843388. www.paynesbeefarm.co.uk

Index

Acarine	67
African bees	90
American Foul Brood	68, 69
Amoeba	69
Ants	71
Artificial swarm	60
BBKA	94
Bee Associations	41
Bee plants	23, 82-5
Bee space	8, 10, 21, 25, 33
Bee stings	90
Bees	
buying	42-3
fanning	45
history	9-11
legislation	91
Braula	70
Brood	
box	29
comb	16, 49-51
frames	32, 34, 46, 48-9, 51
Capping	72, 75, 76
knife	72, 75
Cast	59
Cells	50, 55
Chalk Brood	68, 69
Clothing	36, 37
Colony	42-3
Comb	50
Crown board	31, 48
Disease	47, 60
Drones	16, 17-8, 19, 49, 55, 86
European Fowl Brood	68, 69
Fanning	45
Feeder	38-40, 52-3
Feeding	52, 53-4, 59, 86, 89
Foundation	47
Frames	32-5, 39, 49
Fructose	19
Glucose	19
Heather	77
Hive	
accessories	29, 38-40
moving	43, 64-6
opening up	45, 46-51, 87
siting	44-5
types	24-7, 80-1
weatherproofing	33
Honey	50, 88
bottling	76
buying	41
comb	77
extracting	73-6
flow	22, 23
heather	77
rape	77
sales	78
Larvae	17, 18
Legislation	
bee	91
bee products	78, 91
Mead	79
Mice	71
Moving hives	43, 64-6
Nectar	19-20
Nosema	67
Nucleus	42, 47, 62, 86
Open mesh floor	29
Overcrowding	57
Paralysis	69
Plants	23, 82-5
Poisoning	71
Pollen	20, 49
feeding	54, 86
Pollination	22
Porter bee escape	73
Propolis	21
Queen	15, 16-7, 19, 45, 46, 49, 59
cells	55, 59
excluder	29, 31, 45, 48, 66, 87
marking	61-2
replacement	64
wing clipping	57
Rape	77
Robbing	71
Royal jelly	17
Sac brood	69
Showing bee products	81
Skep	9-10
Smoker	36-8, 39, 48
Solar wax extractor	78-9
Sucrose	19
Super	26, 30-1, 73, 78
Supersedure	61
Swarm	
artificial	60
hiving	56, 58-9
prevention	57
signs	55, 87
taking	58
Uniting colonies	63
Varroa	13, 29, 42, 43, 47, 60, 68, 70, 86, 88
Wasps	71, 88
Water	20
Wax	21
candles	79
extraction	78
moth	71
polish	78
Woodpeckers	71
Workers	16, 18, 19-20